Smarter Solar: Cost-Effective Algorithms for Peak Power Under Any Conditions

Vani

Table of Contents

Chapter 1:   Introduction

In this book an investigation was done to determine the drawbacks of conventional algorithms (if any) for MPPT and how computational intelligent optimisation algorithms can be used to cope with these drawbacks. All the components involved in designing the complete system were investigated and discussed. This includes the static and dynamic behaviour of DC-DC converters, modelling of the PV cell using the single diode model. The unique condition of partial shading was investigated to understand how it alters the PV power –voltage characteristic curve.

1.1 Background

The traditional means of power generation through fossil fuel driven plants has since received severe criticism as it has unsolicited environmental effects on the atmosphere. The use of renewable energy has thus been motivated as a result.

Research on renewable and distributed energy sources is ongoing at accelerated rates all over the world. Photovoltaic (PV), wind and biomass are the most common and successful renewable energy sources currently in use. Grid-connected renewable energy sources like solar energy system uses power electronics converters to step up DC voltages, the renewable energy sources are often tied to the grid allowing power transfer from the inverter to the distribution grid. The interconnection between these small generation systems with the grid has since grown into a network called the Distributed Power Generation System (DPGS). Through the DPGS, renewable energy has since played a pivotal role in today's energy scenario.

Photovoltaic (PV), wind and biomass have continuously been developed and have shown that they are reliable and cost competitive [1]. The cost of these renewable energy sources is currently decreasing, and further decreases are expected with the increase in demand and improvement of production [2]. Many countries have adopted new energy policies to encourage investments in renewable energy sources.

Solar energy can be harvested by means of photovoltaic cells. The major drawback of solar energy is that it is not available all the time and also the solar plants sometimes experience cloud cover problems.

1.2 Motivation

Large Solar plants and residential PV installations are becoming a major source of power generation in South Africa and worldwide. Various methods have been proposed in the literature on how to deal with the partial shading that can occur due to clouds and the mutual shading that occurs between adjacent PV blocks in solar plants. In residential PV installations, partial shading is even common due to buildings, trees, clouds, etc. The main method that exists to deal with partial shading is hardware fixture [3]. This approach is complex and expensive [3]. A simple and inexpensive way to deal with partial shaded conditions is therefore needed. Under partial shaded conditions, the PV power vs voltage characteristic curve has multiple power peaks. Computational intelligent (CI) control algorithms can be used to track the global maximum power point (GMPP). These algorithms provide a cheap method in dealing with partial shading conditions and are less complicated.

The major drawbacks of conventional optimisation algorithms such as PnO and IC are that they cannot distinguish between a global maximum power point (GMPP) and a local maximum power point (LMPP) under partial shading conditions. If the algorithm converges at a LMPP this results in power loss.

1.3 Scope and Limitations

This research is based on using computational intelligence as a tool for maximum power point tracking in PV systems. Optimisation using computational intelligence (CI) requires knowledge of identifying the function to optimise and the number of variables available.

The work covered in this book includes, understanding maximum power point in PV systems and why it's necessary. The converter selected will be the non-isolated boost converter, other non-isolated or isolated converters will not be covered in this work.

Precisely, the work in this research covers the application of the Particle swarm optimisation, Firefly algorithm and Fuzzy logic to MPPT control in PV systems. Validation of the CI algorithms is done by comparing them with widely used conventional algorithms.

1.4 book outline

Chapter 2

Provides a literature review of how the photovoltaic cells convert solar radiation energy to electrical energy. The chapter also shows how these cells are interconnected to create PV modules and then these modules are combined to build PV strings and PV arrays. An overview of the PV models used and the introduction of the functional principles and the electrical characteristic of the PV modules are presented. A review of how these PV modules are affected by the irradiance and temperature is presented. The Chapter also discusses partial shading and its effects on the solar PV output. The mitigation methods that exist to deal with partial shading are briefly discussed.

Chapter 3

A review of the relevant optimisation techniques is discussed. The different types of optimisation methods that exist i.e., deterministic and stochastic based optimisation are discussed. A brief review of examples of deterministic and stochastic based algorithms is undertaken showing the advantages and disadvantages of the techniques.

Chapter 4

Discussed the maximum power point techniques, the principle of maximum power transfer and load matching; the use of classical MPPT techniques is also discussed. The advantages and disadvantages of the conventional methods are explained. The chapter also provides a review of the use of computational intelligent algorithms in MPPT, and discusses their necessity under partial shading conditions. CI methods such as Fuzzy Logic (FL) are discussed. The dynamic behaviour of the converter selected is also studied based on the state space averaging technique.

Chapter 5

This chapter introduces the Firefly (FA) optimisation algorithm based MPPT, the Particle swarm optimisation (PSO) algorithm. These Intelligent algorithms are discussed based on how they reduce oscillations at MPP, increase MPP extraction efficiency and increase tracing speed. It is shown that the PSO and FA are able to track the GMPP for any partial shading pattern.

Chapter 6

The implementation of the complete system in MATLAB/SIMULINK is described. This includes each of the subsystems in detail, the solar cell model, boost converter, dynamic behaviour of the boost converter with DC load and the proposed maximum power-point tracking methods. The MATLAB results of the maximum power point tracking methods applied in photovoltaic are also compared and discussed. Each MPPT method is investigated based on the performance criteria of convergence time, extracted power, and efficiency. The performance under different weather conditions and partial shading is also investigated.

Chapter 7

The chapter presents the conclusions and future work.

1.5   Contributions

The major contributions of the book are

1. A clear illustration of the effect of starting point of deterministic algorithms in a multimodal objective function is illustrated. It is difficult to know where the GMPP will occur in partial shaded conditions hence it was shown that using gradient based algorithms under these conditions is not reliable.

2. The metaheuristic algorithms were shown to perform well for MPPT in PV systems than the PnO, IC and Fuzzy logic. The PSO and FA where able to find the GMPP under partial shading conditions where the deterministic algorithms failed.

Chapter 2: Literature Review

This chapter provides an explanation of solar cells, reviewing the relevant semiconductor physics of the solar cell and the general concept of how it converts solar irradiation into electricity. The effect of direct coupling a resistive load on the solar cell is also reviewed. The effects of partial shading are also discussed and how it is solved in the industry.

2.1 Photovoltaic Modules, Strings and Arrays

2.1.1 The semiconductor p-n junction diode

Figure 2.1 shows the electrical symbol of a diode [1].

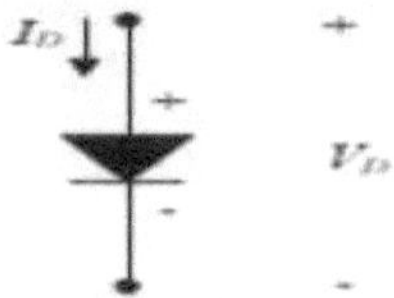

Figure 2.1 Symbol of a p-n junction diode[1]

2.1.2 The Solar cell

Figure 2.2 shows the structure of the PV cell. The solar cell is made of a semiconductor of p-type and boron atoms to form the substrate. A *p-n* junction is created when a high-temperature diffusion process is used to add atoms of phosphorous to the substrate. Where the two semiconductors meet, the holes in the *p* region move into the n region, which leaves behind negative charge ions in the *p*-side. The electrons from the n-side move into the p-side leaving behind positive charged ions in the n-side. Due to the rearrangement of positively and negatively charged ions the depletion region shown in Figure 2.2 is created [1].

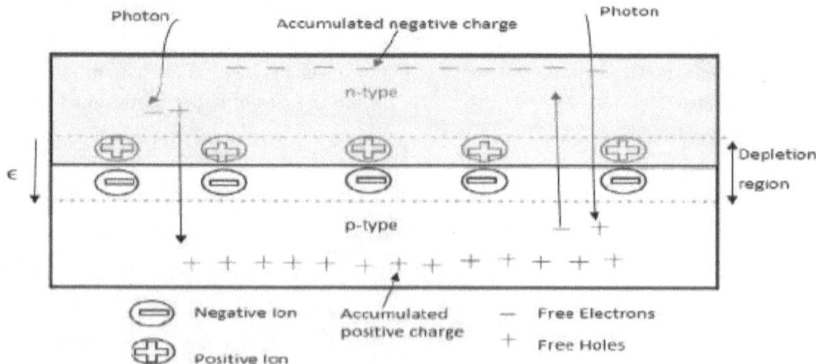

Figure 2.2: The physics of a p-n junction semiconductor[1]

Charge carriers are not found in the depletion region. Hole-electron pair charges will be constantly be created when the p-n material is exposed to solar radiation with enough photon energy. The electrons in the p-n material will be excited because of the photon energy. If a semiconductor with n-type and p-type of a PV cell is connected to an external load as shown in Figure 2.3.

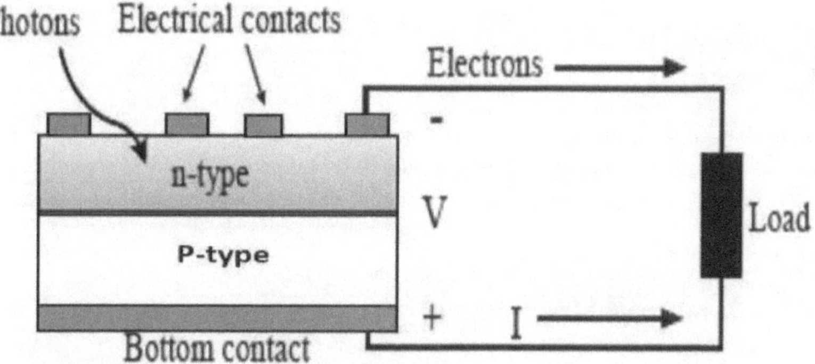

Figure 2.3: A solar cell connected to a load and the flow of current[1]

The electrons in the n-type semiconductor will move through the external load to combine with the holes in the p-type semiconductor. Figure 2.3 also shows how current is produced.

Mono-crystalline and poly-crystalline are the most used materials to make solar cells [1], [2], [3] .

2.1.3   Basic model of a solar cell

There are many different mathematical photovoltaic models that have been developed in the literature. The models are meant to give an estimation of how the PV cell would behave compared to the real cell. The accuracy of the model depends on how many internal variables are considered. A p-n junction diode connected across a current source is usually used to represent a basic solar cell. Figure 2.4 shows this model [4], [5] , [6].

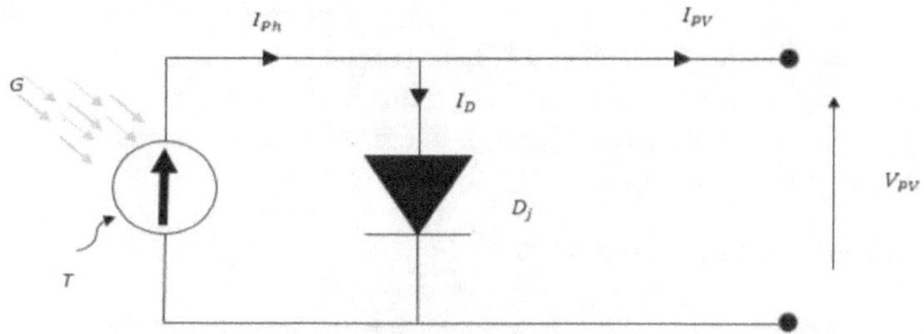

Figure 2.4: Basic PV cell model [4]

In Figure. 2.4, D_j is the ideal p-n diode I_{PV} and V_{PV} are the PV cell current and voltage respectively. The photocurrent produced by sunlight is represented by the current source. Applying the Kirchhoff current law to the model, the current-voltage characteristic function can be obtained as given in equation (2.1):

$$I_{PV} = I_{Ph} - I_D \tag{2.1}$$

where I_D is the current through the diode also known as internal diffusion current and I_{Ph} is the photon current which varies linearly with the sunlight radiation.

The diode internal diffusion current is represented by equation (2.2) below:

$$I_D = I_S \left[\exp\left(\frac{qV_{PV}}{AkT_C}\right) - 1 \right] \tag{2.2}$$

where A is diode ideality factor usually between 1 and 2, k represents the Boltzmann's constant which is 1.38×10^{-23} J/K, T_C is the operating temperature of the cell in kelvin, the charge of an electron q is 1.6×10^{-19}. The cell dark saturation current I_S which changes with temperature is represented by equation (2.3) [7].

$$I_S = I_{RS} \cdot \left(\frac{T_C}{T_{Ref}}\right)^{\left(\frac{3}{A}\right)} \cdot \exp\left[\frac{qE_{gap}}{Ak}\left(\frac{1}{T_{Ref}} - \frac{1}{T_C}\right)\right] \tag{2.3}$$

where I_{RS} is the cell reverse saturation current in ampere at T_{Ref} where T_{Ref} is the solar cell reference temperature in kelvin, usually 298K (25°C). The band-gap energy of the semiconductor material is represented by E_{gap}. Equation (2.4) represents the cells reverse saturation current at reference temperature [7].

$$I_{RS} = \frac{I_{SC}}{\exp\left(\frac{qV_{oc}}{AkT_{Ref}}\right) - 1} \tag{2.4}$$

where V_{OC} is the open-circuit voltage at reference temperature T_{Ref}.

The photocurrent, I_{Ph}, is represented by equation (2.5).

$$I_{Ph} = [I_{SC} + K_1(T_C - T_{Ref})]\frac{G}{G_r} \tag{2.5}$$

where I_{SC} is the short circuit current of the cell, K_1 is the temperature coefficient of the cells short circuit in (amps/K), G is the solar radiation in W/m^2 and G_r represents the solar radiation reference, $G_r = 1kW/m^2$. I_{SC} and V_{OC} are obtained under standard test condition (STC) at a reference temperature of 25 ° C and solar insolation of $1kW/m^2$ [4]- [6].

2.1.4 General model of a solar cell

Figure 2.5 shows the general model that is mostly used in simulation software packages for prediction of solar production. It is more accurate than the model shown in Figure 2.4 this is because it includes the shunt resistance R_{Sh} and series resistance R_S.

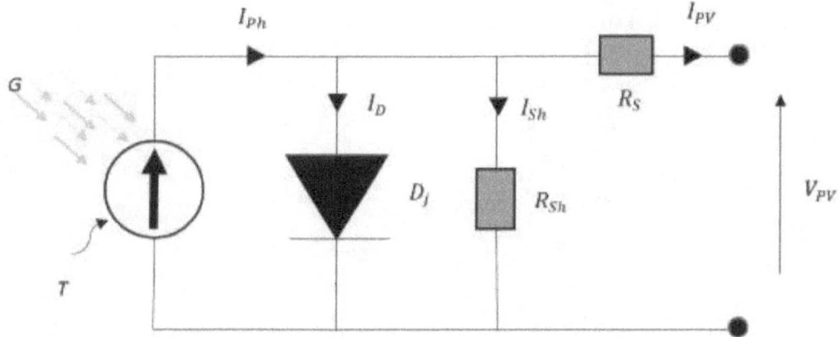

Figure 2.5 General model mostly used in PV cell [5]

The PV cell output current I_{PV} in equation (2.1) becomes

$$I_{PV} = I_{Ph} - I_D - I_{Sh} \tag{2.6}$$

$$I_{PV} = I_{Ph} - I_S\left[\exp\left(\frac{q(V_{PV}+R_S I_{PV})}{AkT_C}\right) - 1\right] - \left(\frac{V_{PV}+R_S.I_{PV}}{R_{Sh}}\right) \tag{2.7}$$

The p-n junction non-idealities and impurities near the junction which causes shunt leakage current to the ground are represented by the shunt resistance R_{Sh}. The bulk resistance of the semiconductor material is represented by R_S. The effects of of R_{Sh} on the PV characteristic curves are shown in Figure 2.6 and Figure 2.7, respectively. From Figure 2.7, it can be seen that the MPP (P_{max}) i.e. the turning point of the curve, increases with the increase of R_{Sh}. The increase of R_{Sh} reaches a point where a further increase has no effect on the maximum power.

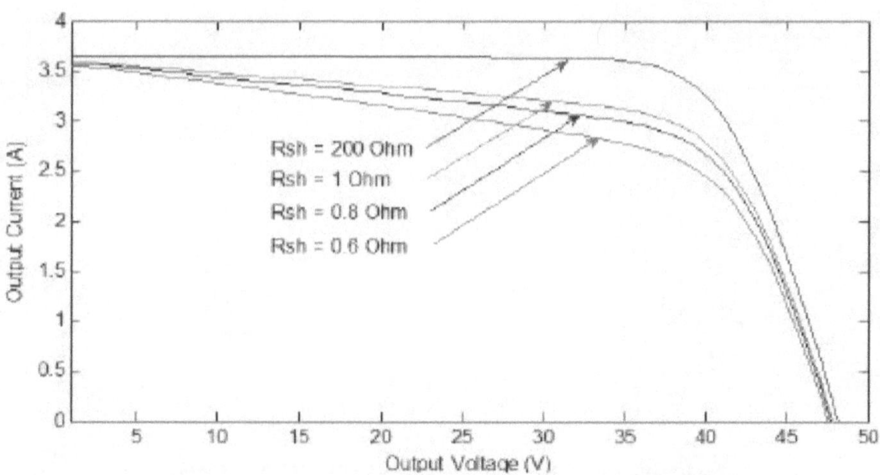

Figure 2.6: The effects of R_{Sh} on the I-V characteristic curve[2]

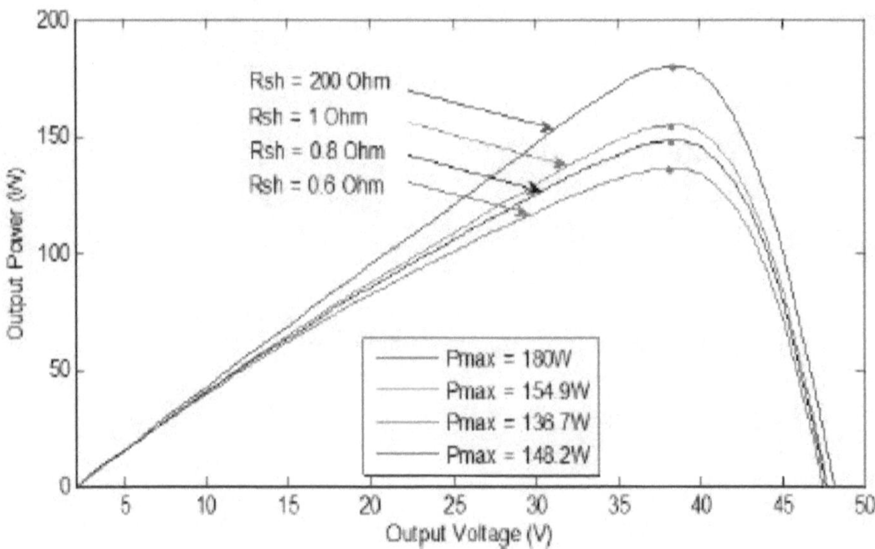

Figure 2.7: The effects of R_{sh} on the P-V characteristic curve[2]

A small variation on R_S results in no change on the I_{SC} and the V_{OC}. The effects on the I-V and P-V characteristic curves are shown in Figure 2.8 and Figure 2.9, respectively [5] , [8]. From Figure 2.9 it can be seen that an increase in R_S reduces the maximum power.

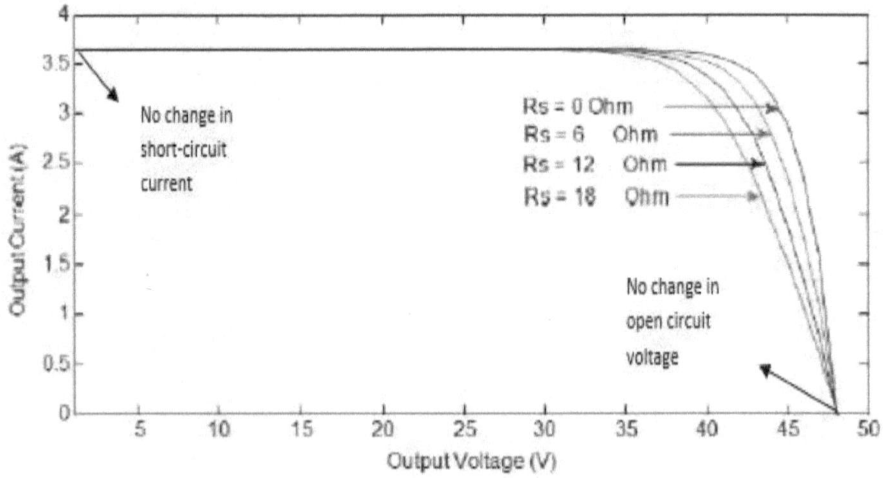

Figure 2.8: The effects of R_S on the I-V characteristic curve[2]

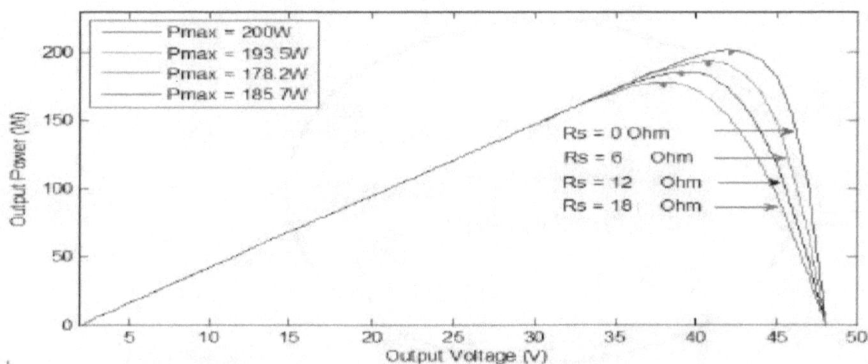

Figure 2.9: The effects of R_S on the P-V characteristic curve[2]

To make the model more compelling and accurate R_{Sh} and R_S are selected so that the calculated maximum power P_{max} is equal to the experimental one $P_{max,exp}$, at standard test conditions (STC) conditions seen on the module datasheet. Iterative processes are used to find the best values of R_{Sh} and R_S. There is only one pair of R_S and R_{Sh} that will produce the required P_{max}. Iterative equations that can be used to find R_{Sh} are shown below [9].

$$I_{mp,ref} = \frac{P_{mp,ref}}{V_{mp,ref}} = \frac{P_{max,exp}}{V_{mp,ref}} \tag{2.8}$$

where $V_{mp,ref}$ is maximum power voltage at STC.

$$I_{mp,ref} = I_{Ph,ref} - I_{S,ref}\left[\exp\left(\frac{q\left(V_{mp,ref} + R_S I_{mp,ref}\right)}{AkT_C}\right) - 1\right] - \left(\frac{V_{mp,ref} + R_S \cdot I_{mp,ref}}{R_{Sh}}\right) \tag{2.9}$$

$$R_{Sh} = \frac{V_{mp,ref} + R_S I_{mp,ref}}{I_{sc,ref} - I_{sc,ref}\{\exp\left[\frac{V_{mp,ref} + R_S I_{mp,ref} - V_{oc,ref}}{a}\right]\}} + I_{sc,ref}\{\exp(\frac{-V_{oc,ref}}{a})\} - (P_{max,ex}/V_{mp,ref})} \tag{2.10}$$

$$a = \frac{N_S AkT_C}{q} \tag{2.11}$$

where N_s is the number of solar cells connected in series. T_C is temperature.

The iteration starts at $R_S = 0$ which must be increased in order to move the modelled MPP until it equates with the experimental MPP [9]. The corresponding R_{Sh} is then computed. Figure 2.10 shows the flow chart.

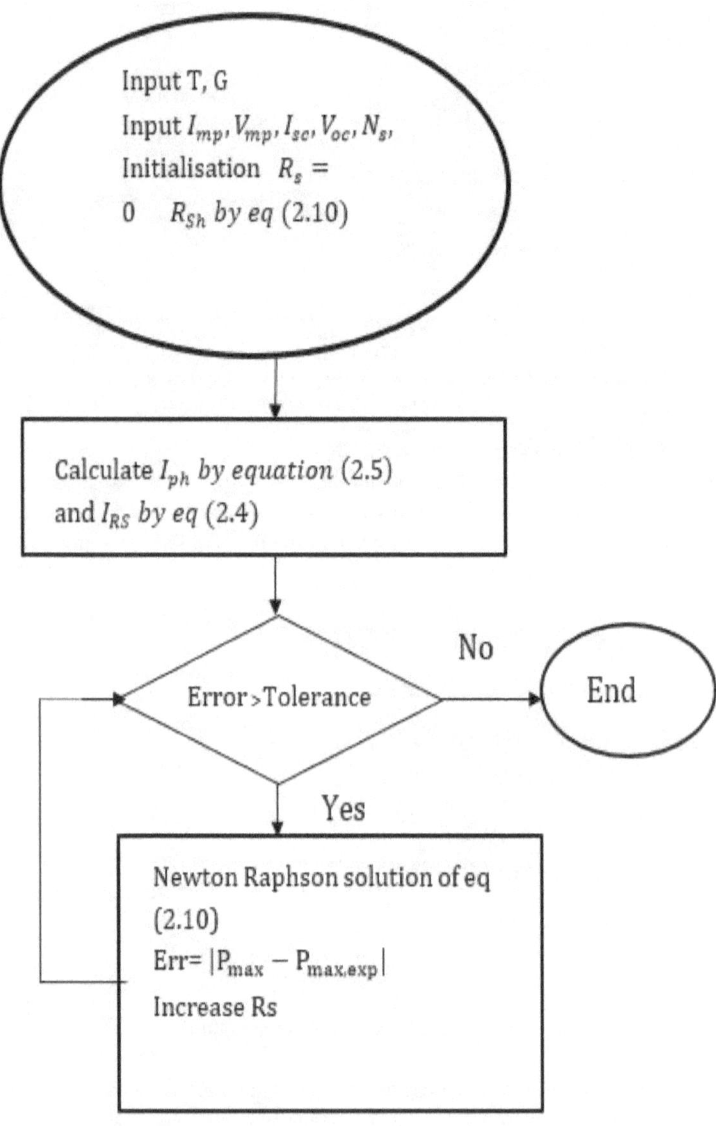

Figure 2.10 Iterative process of finding R_{Sh} and R_S [9]

2.1.5   PV module and array

Because the solar cell power is very small around 2W at 0.5V, the solar cells are connected in series and parallel so as to form a module to produce a desired output power and voltage.

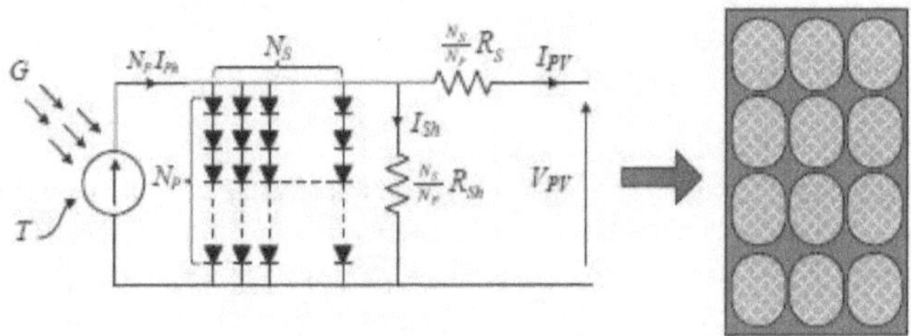

Figure 2.11: Solar cells interconnected to form a module[6]

The same current flows in two or more PV modules connected in series and the total output voltage is the sum of each PV module voltage connected. Hence equation (2.7) can be written as shown in equation (2.12). The PV module current ads up for each module connected in parallel, and the output voltage across remains constant.

$$I_{PV} =. N_p I_{Ph} - N_p I_S \left[\exp\left(\frac{q\left(V_{PV} + \frac{N_S}{N_P} R_S I_{PV}\right)}{N_S A k T_C}\right) - 1\right] - \left(\frac{V_{PV} + \frac{N_S}{N_P} R_S . I_{PV}}{\frac{N_S}{N_P} R_{Sh}}\right) \qquad (2.12)$$

where N_P are cells in parallel and N_s is the number of PV cells connected in series. If these PV modules are connected in series they form a string where the current in the string stays the same but the voltage is a multiple of the number of modules in the string. When the PV strings are connected in parallel they form a PV array in this case the current flowing in one string is a multiple of the number of strings in the array and the string voltage is the same as the array voltage.

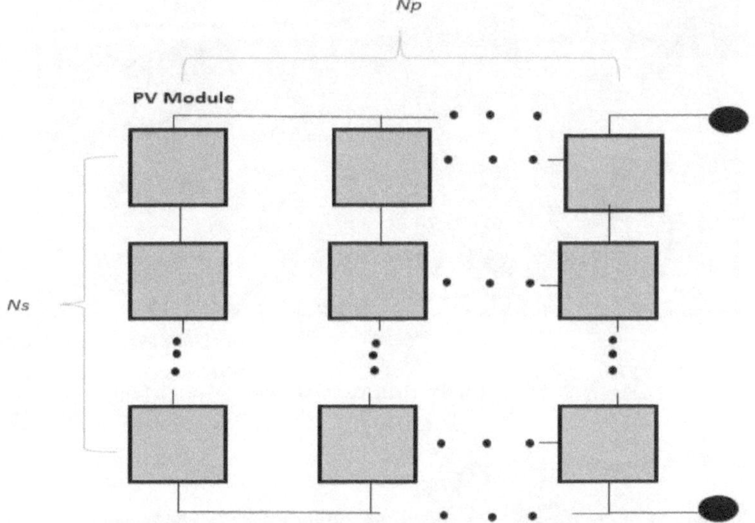

Figure 2.12: Modules interconnected to foam a PV array

2.2 Characteristics Curves of PV

2.2.1 I-V Characteristics

I_{SC} and V_{OC} describe the electrical performance of a PV cell. The point where the curve crosses the vertical axis is the short circuit current. It is considered as the maximum possible current in the circuit.
Open circuit voltage is the point where the curve intercepts with the horizontal axis. It is considered as the maximum possible output voltage the circuit can produce.

2.2.2 Fill factor (*FF*)

The ratio of output PV power at MPP to the power result from multiplying V_{OC} by I_{SC} is called the fill factor of the PV cell. Equation (2.13) shows this. The shape of the photovoltaic cell characteristic depends on these fill factor as shown in Figure 2.13. A high quality cell with low internal losses has a high fill factor [9]. Equation (2.13) can be further simplified to equation (2.14).

$$FF = \frac{I_{MPP}V_{MPP}}{I_{SC}V_{OC}} = \frac{Area\ B}{Area\ A} \tag{2.13}$$

$$I_{SC}V_{OC}FF = I_{MPP}V_{MPP} = P_{max} \tag{2.14}$$

where I_{MPP} is the current at MPP and V_{MPP} is the voltage at MPP. The fill factor depends on the material used and is always less than one. A better operation performance of the PV cell is experienced if the fill factor is closer to unity. The fill factor is affected by R_{Sh} and R_S as shown in Figure 2.6 and Figure 2.8 [8].

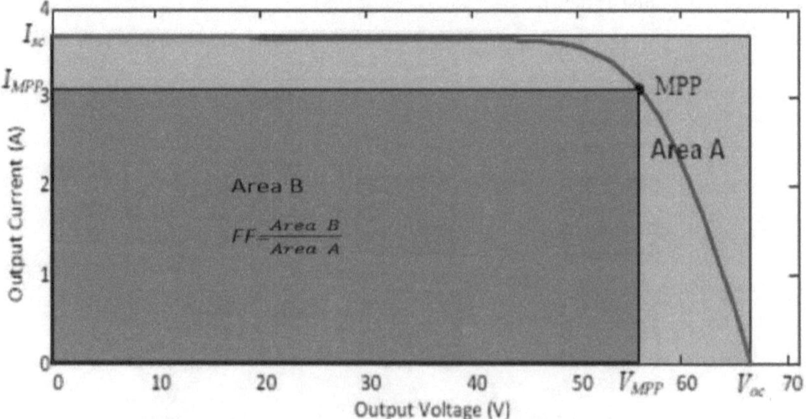

Figure 2.13: I-V characteristic curve showing fill factor[8]

2.2.3 The effects of temperature

Temperature is one of the parameter that affects the output power of photovoltaic panels. Figure 2.14 shows the effect of temperature at constant irradiance.

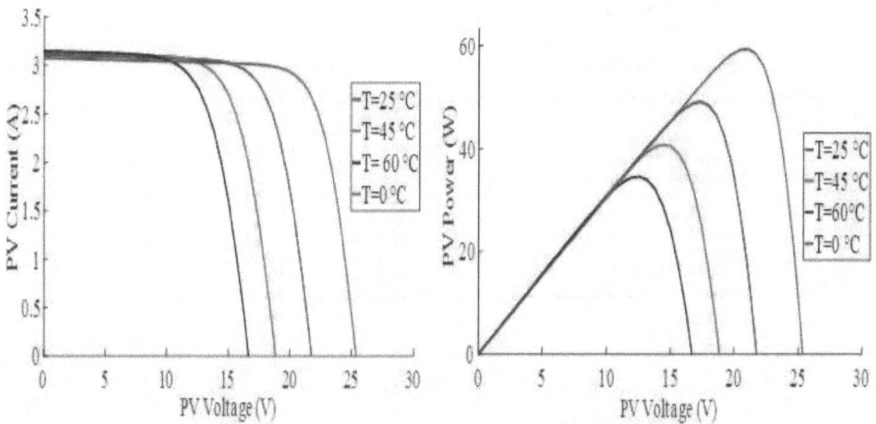

Figure 2.14:The effects of temperature on the photovoltaic I-V and P-V characteristic [10]

As can be seen from Figure 2.14, temperature affects mostly the open circuit voltage. It is clear that an increase in temperature results in a decrease in PV output power and a decrease in temperature results in an increase in PV power [10].

2.2.4   The effects of irradiation

Irradiation is another parameter that affects PV output maximum power as shown in Figure 2.15. From Figure 2.15, it is clear that as the irradiance increases the short current increases. The increase in irradiance results in an increase of PV output power. The short circuit current depends totally on irradiance and it varies linearly with it [10].

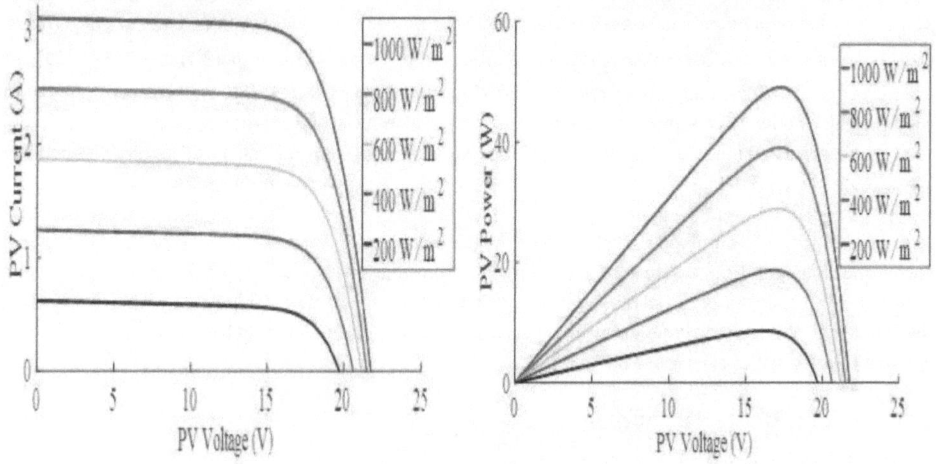

Figure 2.15:The effects of irradiance on the photovoltaic I-V and P-V characteristic[10]

2.2.5 Solar I-V characteristics with resistive load

The PV cell will function in two main operating characteristic regions these are the voltage source region and current source region. Figure 2.16 shows the locations of these regions.

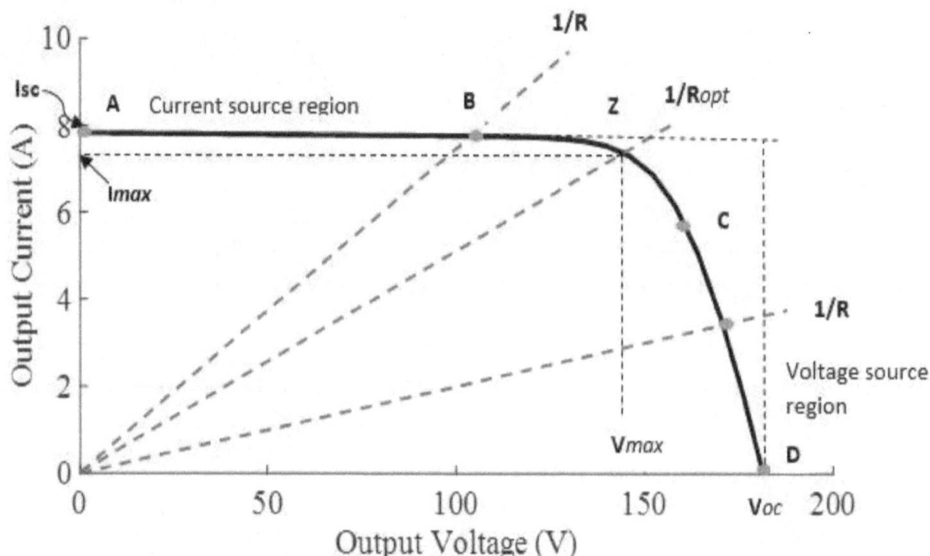

Figure 2.16: The intecsection of the resistive load curve with the photovoltaic I-V curve[11]

When a constant resistive load (R) is coupled directly to a PV module terminals, the operating point of the connection will depend on where the intersection of the PV cell curve with the load curve occurs. This can be seen in Figure 2.16. The resistance load has a straight line characteristic with a slope, $\frac{I}{V} = 1/R$ [11],[12]. The size of the resistive load determines the power transferred from the PV source. If the resistance of the load is small, the PV cell operates in the current source region AB of the characteristic curve. When a large load resistance is connected, the PV cell operates on the voltage source region CD of the characteristic curve [13].

It is clear that the operating point of the load might not be at Z which is the MPP of the PV array and furthermore as seen from Figure 2.14 and Figure 2.15, this maximum power point will constantly vary with environmental changes of temperature, solar irradiance and really gets complicated under partial shading were multiple points of maximum power will exist. This variation is nonlinear which makes the matching of the two characteristics even more difficult [12].

2.3 Effects of partial shading

It was seen that under a constant irradiance, only one maximum power point exists. In reality, the irradiance exposed on modules connected in an array rarely is the same. This can be due to cloud cover, daily sun angle changes, shading from adjacent PV modules or trees and buildings [14]. Power is lost due to shading, current mismatch in a string and voltage mismatch between strings in parallel. When PV panels connected in series do not receive the same solar irradiance partial shading will occur [15].

2.3.1   Partial shading and the bypass diode effects

As mentioned previously current that flows in modules connected in a series is the same including those which are under a shade. The shaded cells can act as loads if they get reverse biased this will cause them to consume power produced from the fully irradiated cells. Problems like hot spot occur if the modules are not protected. A bypass diode is connected across each module to mitigate this problem.

During normal operation of the panels without shade, the bypassed diode is reverse biased and has high impedance. Under shaded conditions, the bypass diode across the shaded module terminal is in forward biased, therefore it conducts the current produced by the unshaded modules. Since the shaded modules are bypassed, multiple peaks on the P-V curve and multiple stairs on the I-V curve are exhibited. Figure 2.17 shows how the bypass diode is connected [15].

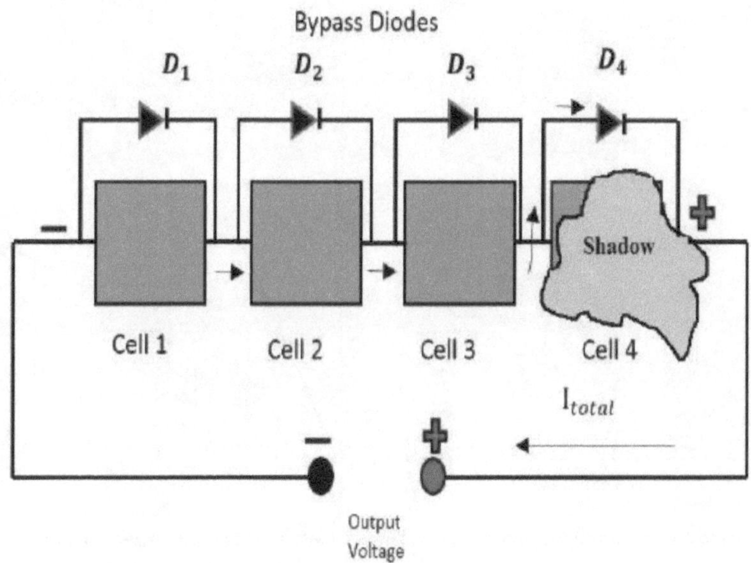

Figure 2.17: Connection of Bypass Diode across the module[15]

2.3.2   Characteristic curves under partial shading

The bypass diodes connection will change the uniform PV characteristics curves of the panel, resulting in multiple peaks [16], [17]. Figure 2.18 illustrates this point.

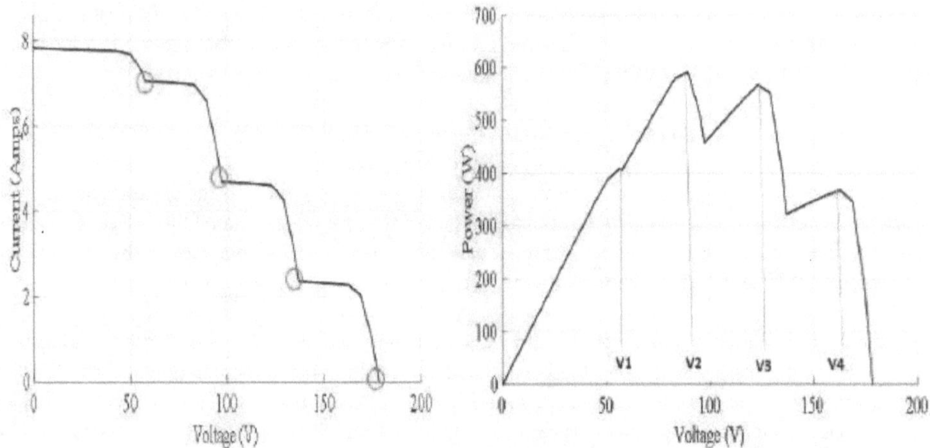

Figure 2.18: I-V and P-V characteristics under partial shading[15]

Figure 2.18 shows the different voltage values were the different power peaks will occur. It can be seen that the GMP occurs at low voltage (V2). Figure 2.19 shows the GMP ocuring at a midium voltage on the P-V characteristic curve.

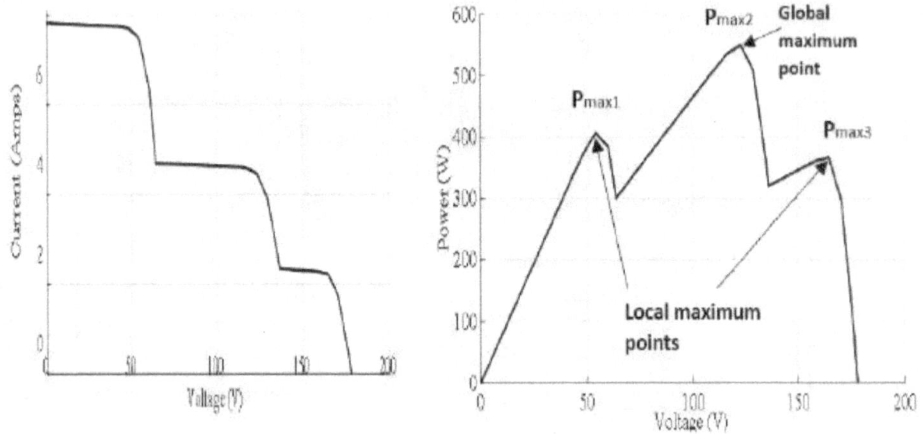

Figure 2.19: : I-V and P-V characteristics under PSC showing local power and global[15]

It can be observed that the P-V curve has multiple maximum power peaks [18]. The position of the GMPP on P-V curves is not fixed as illustrated in Figure 2.19 and Figure 2.18. From Figure 2.18 and Figure 2.19 it can be seen that the number of multiple power peaks is equal to the number of stairs on I-V characteristic.

2.3.3 Mitigation methods for partial shading

Two methods are generally used to mitigate the shading effect. The first is based on hardware fixtures [19], [20], [21] , [22] ; multilevel converter system [23], allowing each PV source connected in series to be controlled separately to its MPP; and power electronic equalizers [24]. This approach is complex and costly [25] hence an alternative cheaper method has to be found.

The second approach is to track the global maximum power (GMP) by developing computational intelligence algorithms (CI), and this will be the focus of this book. Computational intelligent algorithms have been suggested to solve the multiple power peak problem in solar PV systems. The multiple power peaks can be seen as a stochastic optimisation problem and global optimisation algorithm techniques can be used to find the global peak (best power).

2.4   DC-DC Boost Converter

The switch mode DC-DC converter is a critical component for MPPT system. It is responsible for maximum power transfer from the PV source to the load. It achieves this by regulating the PV input voltage and producing a controlled dc output voltage [26].

A direct connection of a PV panel with a resistive load will result in the load operating where the characteristics of the curves intersect, this could be the MPP or not. The converters ensure that the PV array is forced to operate at the MPP. MPPT algorithms are used to control the duty cycle. This control of the duty cycle allows the curves to be matched at MPP even with the nonlinearly changes of the PV array power due temperature and irradiance. Different loads will have their own unique characteristic curves [2].

They are two types of DC-DC converter topologies these are isolated and non-isolated converters. Isolated converters use a transformer to isolate the input from the output [27]. They are commonly used in switched mode power supply [27], [26]. Non-isolated converters are the boost and buck converters. In this book, only the boost converter topology will be reviewed.

The boost converter output voltage is always more than the input voltage [26]. Figure 2.20 shows the circuit topology of the boost converter.

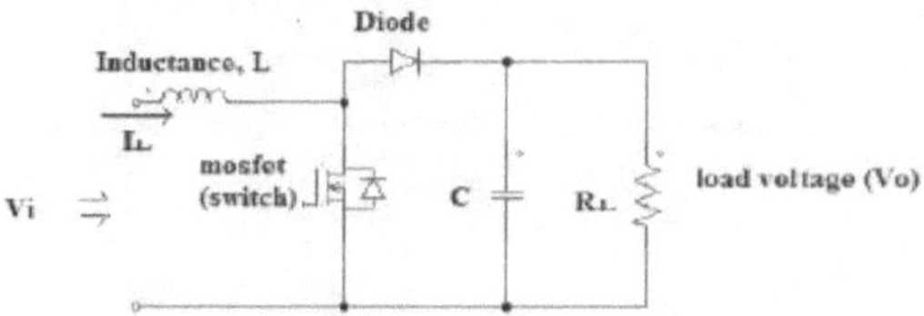

Figure 2.20:DC-DC Boost Converter [26]

The boost converter consists of four elements. These are inductor, diode, capacitor, and the MOSFET. The Boost converter can be used as a switching-mode regulator. The regulation is usually accomplished by pulse width modulation technique (PWM). The converter is operated in continuous conduction mode. Figure 2.21 shows how the voltage and inductor current behaves in this mode [26].

$$Vd_{ton} + (Vd - Vo)_{toff} = 0 \qquad\qquad (2.15)$$

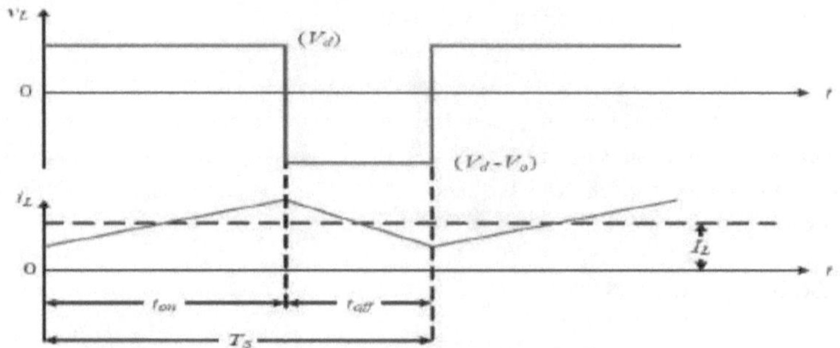

Figure 2.21 :How the voltage and inductor current behaves in   continuous current mode [26].

Figure 2.22 shows the two operating states of the boost converter [26].

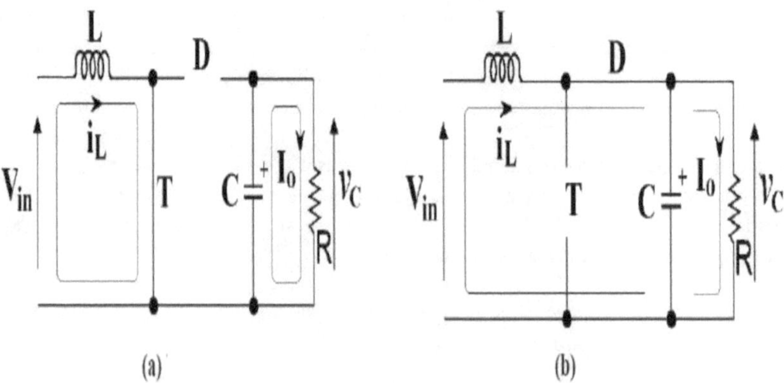

Figure 2.22 Equivalent Circuit for boost converter (a) switch ON; (b) switch OFF [26].

Equation (2.15) can be evaluated to obtain the steady state equation (2.16) [26]

$$\frac{V_o}{V_d} = \frac{T_s}{t_{on}} = \frac{1}{1-D} \tag{2.16}$$

Where D is the duty cycle.

Assuming no losses in the circuit, power input is the same as power output [26].

$$P_d = P_o$$
Therefore
$$V_d I_d = V_o I_o$$
This results in

$$\frac{I_o}{I_d} = (1 - D) \tag{2.17}$$

Equation (2.18) express the duty ratio D when the boost converter is at a stable state

$$D = 1 - \frac{V_d}{V_o} \tag{2.18}$$

where V_d represents input voltage and V_o output voltages of the converter. The filter inductor and capacitor can be calculated using equations (2.19) and (2.20).

$$C_{out} \geq \frac{I_{out}D}{f_{sw}\Delta V_{out}} \tag{2.19}$$

$$L_{critical} \geq \frac{V_{in}D}{f_{sw}\Delta I_L} \tag{2.20}$$

where f_{sw} is the switching frequency, ΔI_L is the input current ripples and ΔV_{out} is the voltage output ripple

2.5 Summary

This chapter reviews the physical structure of solar cells and explains how electricity is produced when the semiconductor device is exposed to solar radiation. The voltage produced by the solar cell is very small therefore a need to connect the PV cells in a series configuration is required.

The chapter also presents two commonly used PV models for simulation prediction, the single diode model with series resistance and shut resistance and a single diode model without. The single diode model with series resistance and shut resistance is found to be more accurate to represent the actual solar cell and it is the one that is used in this book. The series and shut resistance parameters of the single diode model have to be correctly found to accurately model the actual PV cell. These parameters can be found by iterative methods.

The chapter also discusses how temperature, radiation and the load type used affect the PV power produced. In the literature, it is found that an increase in temperature reduces the PV output power and a decrease in temperature increases PV output power. The increases in radiation increases the PV output power and the decreases of radiation decreases the PV power. The intersection of the characteristic curve of the load and the characteristic curve of the PV current - voltage is where the PV cell will operate. The constant variation of PV power due to changing climatic conditions makes the load matching difficult as a result a MPPT algorithm is needed to constantly extract the maximum power.

Constant radiation or temperature results in one maximum power point on the PV power-voltage characteristic curve but different illumination of radiation on different PV modules connected in series results in multiple maximum power points (partial shading). A bypass diode is connected across each module to prevent hot spots. Mitigation methods are discussed to reduce the effect of partial shading and it was found that most of them are complex and expensive to implement. It is suggested that using computational intelligence algorithms to extract the optimum power from a partial shaded PV array can be less complex and yet still be effective. A review of the DC-DC boost converter is discussed as a critical component in MPPT. The converter is used to match the PV source with the load.

Chapter 3: Optimisation basics

In this chapter a review of relevant literature of optimisation techniques is presented. This includes the deterministic and stochastic optimisation methods. Advantages and disadvantages of these techniques are also discussed.

3.1 Optimisation techniques

Optimization is used in multiple fields from engineering design, computer science etc. . . . The need to maximize or minimize something is always necessary [28]. Optimization is basically searching for the best variables in a function so as to have the best outcome. Thus a problem can only be optimized if it can be expressed mathematically. The decision variables can be either discrete, continuous or a mixture of the two. The search space is the area covered by the decision variables and the space formed by the cost function variables is called the solution space [28]. Optimisation can be done to find the maximum or minimum value of the function. There are two optimization techniques which are classified as deterministic approach and stochastic approach.

3.1.1 Deterministic algorithms

Deterministic algorithms use one solution at a time, which will trace out a path as the iterations continue [29]. Good examples of deterministic approaches are the Hill-Climbing and the Perturb and Observe method. If the algorithm is made to start at the same starting point it will repeat the same path regardless of whether the code is run today or another time. Most conventional algorithms are deterministic [29]. Some deterministic optimization algorithms that use the gradient information are called gradient-based algorithms. An example is the Newton Raphson algorithm it uses the function derivatives to find the optimum point. It works very well for continuous unimodal problems [28]. If the function is now multimodal (having multiple peaks), finding the optimum point might be difficult using gradient based algorithms. For objective functions which are multimodal such as a sine function, gradient based optimization methods are very sensitive to the starting point. If the starting point is far from the sought minimum or maximum, the algorithm will usually get stuck in a local minimum and/or simply fail [28]. Nonlinearity and multimodality are the main problems, which render most conventional methods such as the hill-climbing method inefficient and lead them to be stuck to local inferior solutions [29]. Another challenge that comes about using deterministic approaches is when the number of decision variables are large for example 50000, coupled to the non-linearity of the function it can be impractical to search the number of possible combinations of the different variables that will give the best output. In this case an algorithm that does not use the gradient is preferred. Non-gradient algorithms use the function values and avoid any use of its derivative [29]. Stochastic algorithms also known as heuristic and metaheuristic algorithms are gradient-free algorithms that are designed to deal with these types of problems [28].

3.1.2 Stochastic Algorithms

Stochastic algorithms always have some form of randomness. A good example are the Genetic algorithms (GAs). The solutions in the population will be different each time the program is run this is because the algorithm will use some foam of pseudo-random values. Though the final outcome of the search process may be similar [29], the movement traced by each individual will not be exactly repeatable. Metaheuristic algorithms are also stochastic by nature. Heuristic means to find or discover by trial and error [29]. Meta means beyond or high level. All metaheuristic algorithms rely on a tradeoff of randomization and local search. Randomization allows the search process to be biased on the global scale instead of the local scale [29]. Therefore, almost all metaheuristic algorithms intend to be suitable for global search optimization. In complex optimization problems, quality solutions can be obtained in a reasonable duration of time, but there is no guarantee that the best solutions are obtained [28]. This allows us to find easily obtainable good solutions which are not necessarily the optimum solutions which still allow us to fix the problem. Among the quality solutions found it is assumed that some of them are nearly optimum. It will be difficult to search every possible solution or combination to a given complex problem; the aim is to obtain quality feasible solutions in an acceptable timeframe. Metaheuristic algorithms use two major principles to search in a given search space these are intensification and diversification [28]. Diversification means different multiple solutions will be created so as to explore the global scale. Intensification means to concentrate the search in a local region by analyzing the information that a current quality solution is located in this region [29]. The selection of the best solution ensures that the solutions will converge to the optimality, while the randomization of solutions prevents the solutions from being trapped at local optima and increases the diversity of the solutions [30]. The good combination of diversification and intensification will encourage the global best to be found.

Metaheuristic algorithms can be classified as population-based meaning that one has multiple agents looking for the optimum solution. Examples of metaheuristic algorithms include the Particle Swarm Optimisation (PSO), Firefly (FA), Bat (BA), Harmony search (HS), Cuckoo search (CS), Grey Wolf (GW), Ant Colony Optimisation (ACO), Flower Pollination (FP), etc. Alternatively, there is no universally better algorithms that exist. The main research goal in optimization is to come up with the most suitable and efficient algorithms for a given optimization problem. The complexity of the objective functions usually indicates the complexity of an optimization problem [28].

3.1.3   Optimisation classifications

The classification of optimisation problems based on the number of objectives will result in two categories: single objective and multi-objective. Most real-world optimisation problems are multi-objective [28].

We can also classify optimisation in terms of number of constraints or boundaries. This is basically the range or area where optimisation needs to take place.

We can classify optimisation in terms of the landscape of the objective functions. If there is only a single peak which will be the global optimum, then the optimisation task is unimodal. However most objective functions are multimodal functions where multiple local peaks exist and only one global is required, these are much more difficult to solve for example equation (3.1) is multimodal. It has two variables in the x and y dimension.

$$f(x,y) = \sin(x)\sin(y) \tag{3.1}$$

The design optimization problem variables can be either discrete or continuous or a mixture of both [28].

3.2   Summary

The chapter reviews the basic understanding of optimising functions to find the maximum or minimum point. Optimisation being defined as finding the best variable in a function to produce the optimum output. The optimised function is called the objective function. The area covered by the decision variables is called the search space. Two techniques are used to obtain this optimum variable these are deterministic and stochastic. The deterministic approach uses one solution at a time to trace a path as the iteration continues whilst the stochastic approach uses multiple solutions randomised in a search space. Deterministic techniques are usually gradient based meaning they use the gradient of the function thereby sensitive to where they start. Stochastic techniques uses the values of the function and not the gradient and are given a search space where the multiple solutions can search. The landscape of the objective function can be unimodal meaning having one optimum peak or multimodal having multiple peaks and one optimum peak.

In finding the maximum or minimum point the deterministic approach can get stuck at a point that is not the best (local point) so it is suggested that they should not be used in a function where there multiple maximum or minimum points (multimodal function). The stochastic approach has the ability to search for the best solution where multiple peaks occur in a function without getting stuck at a local point.

In the literature it is found that to best maximise the stochastic approach diversification of solutions and intensification is required.

An optimization problem can have multiple objectives to optimize this is called a multi objective function. Examples of deterministic approach methods are Hill climbing and Newton Raphson and examples of stochastic methods are the Particle Swarm Optimisation (PSO), Firefly (FA), Ant Colony Optimisation (ACO), Bat (BA), Harmony search (HS), Cuckoo search (CS), Grey Wolf (GW), Flower Pollination (FP), etc. Figure 3.1 further shows the summary of optimisation classifications.

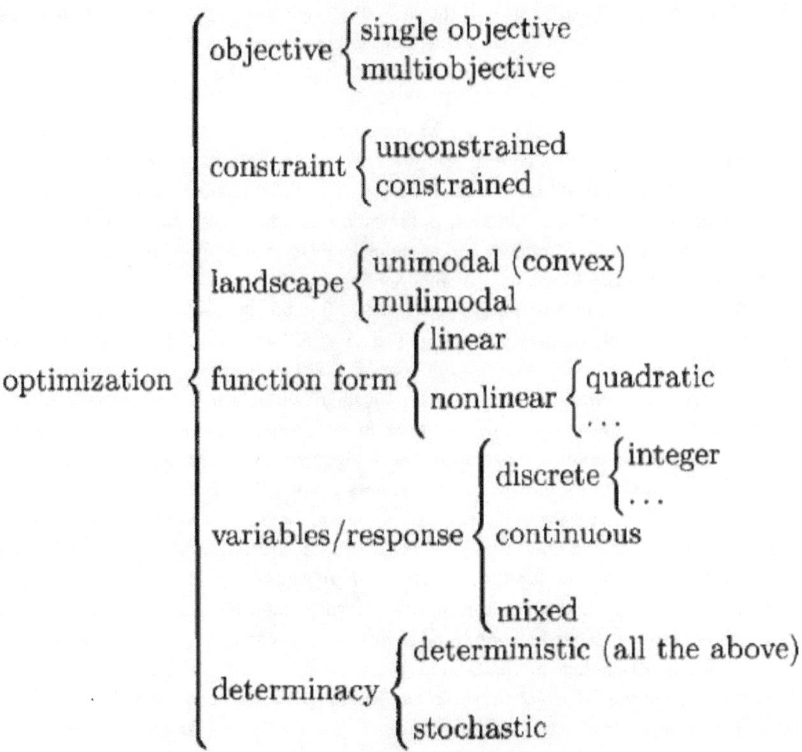

Figure 3.1: Optimisation Classifications [28]

Chapter 4. Maximum power point tracking and DC-DC Boost converter control

This chapter discuses maximum power point tracking in solar PV systems. A clear explanation of why MPPT is necessary under static, varying and partial shading weather conditions in solar PV systems is provided. Different types of optimisation algorithms for MPPT are explained and their advantages and disadvantages are discussed. The DC-DC boost converter dynamic behaviour is also discussed in this section.

4.1   Photovoltaic maximum power point tracking

The conversion efficiency of PV modules is not very high. The efficiency of converting sunlight energy to electrical energy can range from 12-22%. The range can drop even further during partial shading, variation in radiation or temperature and also load changes [31]. As discussed in chapter 2, the characteristic impedance of the load will hardly match the characteristic impedance of the PV system for maximum power transfer to occur so if a load is directly connected to the PV array, maximum power is hardly achieved. It is critical to operate the PV array at the MPP, or as near to it as possible. A load matching circuit can be used to achieve MPPT. A drastic improvement in the power extracted is obtained when using a load matching circuit as compared to when a direct load connection is used [32].

A typical electronic load matching circuit consist of the PV source with a DC-DC converter transferring PV power to the load. To enable MPPT a control algorithm is used to control the duty ratio of the converter [33]. In the literature different MPP techniques for photovoltaic systems are compared [32], [33], [34] , [35], [36]. As discussed in chapter 3, different optimisation algorithms can be used depending on the optimisation problem to find the optimum power point in PV systems. These methods are different in complexity, range of effectiveness, cost, convergence speed, etc.

In ref [37], [38] twenty unique techniques are compared to find out their advantages and disadvantages. The performance of these techniques were summarised so as to know which MPPT technique should be selected for a particular situation. As the problem of MPPT is seen as an optimisation problem, deterministic or stochastic optimisation can be used. A well-known deterministic approach in MPPT is hill climbing method [34]. Then there are many variations of hill climbing commonly used in MPPT like the Perturb and observe and incremental inductance. These algorithms will be discussed further in the next section. Other classical MPPT algorithms were implemented in Ref [35], [36]. Computational intelligence algorithms include neural networks with its hybrids [39] and fuzzy logic control [37]. Metaheuristic based algorithms are also now popular in MPPT these include the Ant Bee colony optimisation [17], Bat algorithm optimisation [40], Grey Wolf optimisation [41], Particle swarm optimisation [42], Firefly optimisation [43] etc. Basically, any optimisation technique can be used to solve the MPPT problem. The problem arises when the landscape of the objective function to be optimise starts changing such as under partial shading conditions. Clearly using hill climbing methods under these conditions would not be reliable because of the multimodality that exist in the P-V characteristic curve. Ideally, using stochastic based algorithms would be best because of their ability to search globally. Figure 4.1 shows a simple maximum power point tracking for PV systems.

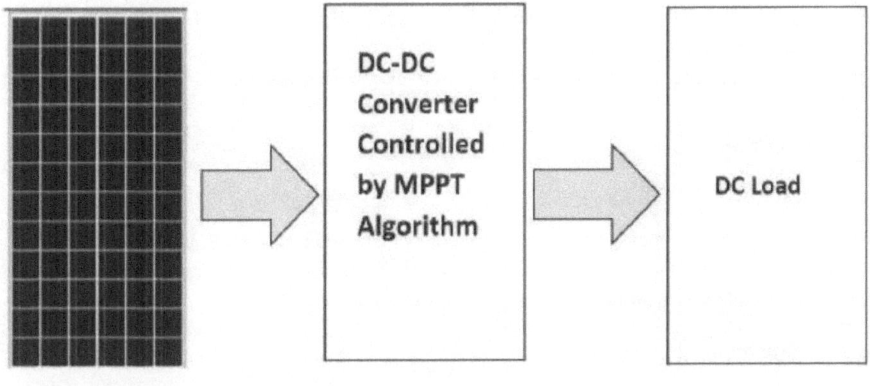

Figure 4.1: Block diagram of a MPPT for PV systems

It is important to make sure that the system is transferring the maximum power regardless of change in load and atmospheric conditions. This is achieved by using the theorem of maximum power transfer where the maximum power is transferred from the source to the load by making the source impedance equal to the load impedance [44].

The MPPT algorithm or optimisation calculating model is supposed to find the MPP even with the nonlinear, unpredictable changes that occur due to variation of temperature and irradiance.

4.1.1 Performance specifications of MPPT control algorithm

For a successful performance design and evaluation of the MPPT control algorithms, performance criteria's are considered [45].

4.1.1.1 Steady-state error

When the maximum power point is obtained the algorithm should stop tracking and force the system to operate at this MPP. This can be an impossible task to achieve in a real world MPPT system because of the constant fixed step size perturbation process in conventional MPPT algorithms like the PnO and IC. Metaheuristic algorithms try to reduce this steady state error, for example the PSO reduces the velocity close to zero when it converges to a value this results in very minimum oscillations at steady state.

4.1.1.2 Dynamic response

MPPT control algorithms need to be fast in tracking the MPP in the rapid changes of climatic conditions. The faster the tracking speed of the MPPT algorithm the more the solar energy is utilised.

4.1.1.3 Tracking efficiency

To quantify how successful a MPPT algorithm is in tracking the MPP and to what extent is it better in extracting the maximum power of the PV system compared to other optimisation algorithms, tracking efficiency is calculated. References [46], [47] defined tracking efficiency as the ratio between the actual power of the PV array tracked by the algorithm and the theoretical power during the same time period.

Due to the wide variation of climatic conditions, different algorithms should be evaluated over a range of different operating conditions to see which performs better. A well designed MPPT should perform well under different atmospheric conditions.

Equation (4.1) can be used to calculate the PV system tracking efficiency [48]

$$\eta_{MPPT} = \frac{1}{n} \sum_{i}^{n} \frac{P_{actual,i}}{P_{max,i}} \tag{4.1}$$

where $P_{actual,i}$, is the ith sample of the PV system power extracted by the MPPT control algorithm, $P_{max,i}$ is the ith sample of data sheet maximum power (theoretical power) of the PV system power that could be produced under the given solar radiation and temperature and n is the total number of samples.

4.2　MPPT algorithms

4.2.1　Perturb and Observe

The Perturb and Observe (PnO) method is generally the most used because it is simple to implement [49], [50], [51], [52]. This method works by increasing the system PV voltage and observing the impact on the output PV power. Figure 4.2 shows the operating principle of the PnO.

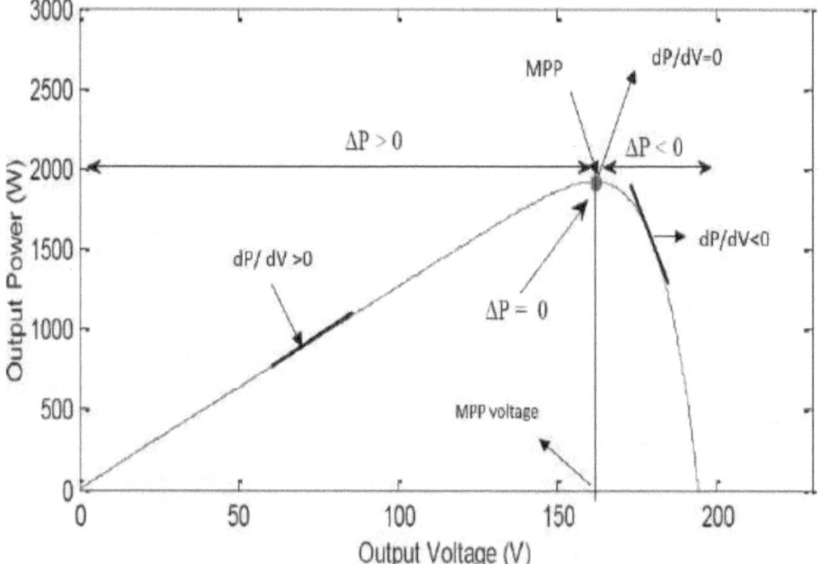

Figure 4.2: How the PnO operates on the photovoltaic P-V curve[49]

The algorithm is summarized in Table 4.1 [50] and the flowchart in Figure 4.3 [49].

Table 4. 1 Algorithm Movement[49]

Sign of change of voltage (dV)	Change in Power (dP)	Direction of next perturbation
positive	positive	positive
positive	negative	negative
negative	positive	negative
negative	negative	positive

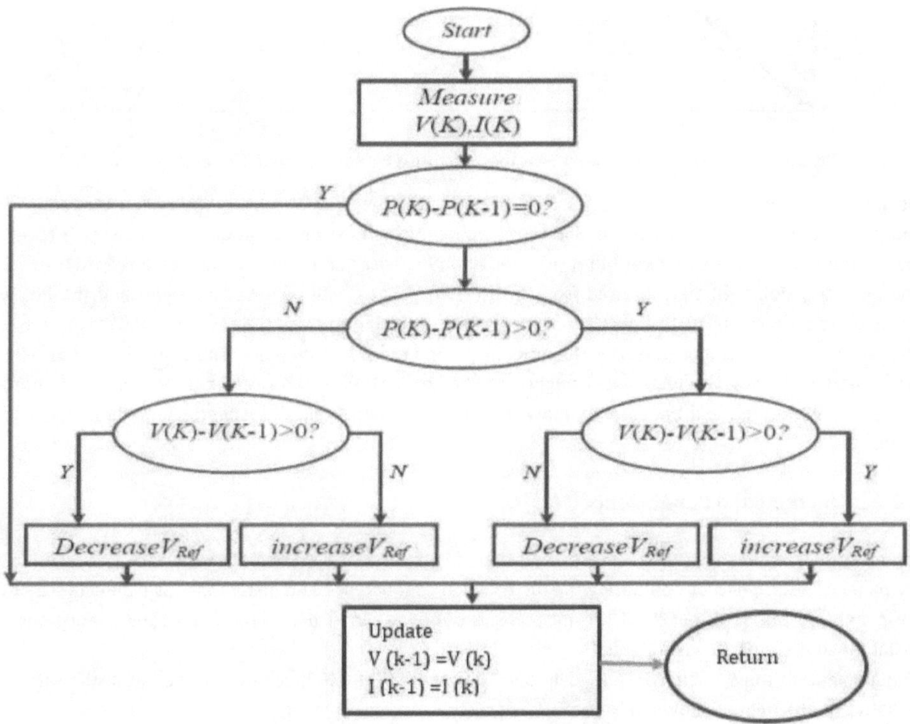

Figure 4.3: Flow Chart of the PnO algorithm[53]

where Vref equals to maximum power point voltage found (Vmpp) [53]. I(k) and I(k-1) is the present PV current samples and previous PV current samples, respectively.

The procedure is repeated until the MPP is obtained. The constant step size in the algorithm causes oscillations around the MPP. If the constant step size of the perturbation is reduced this can result in minimum oscillations. However, a too small perturbation slows down considerably the tracking of the MPP [51], [53]. A trade-off is needed between accuracy and speed [54]. Another drawback of the PnO is that it can fail to track the MPP during fast changing atmospheric conditions such as shown in Figure 4.4 [37], [49].

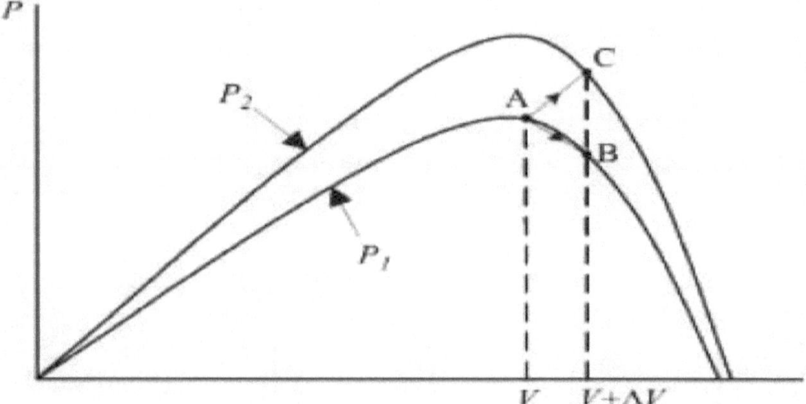

Figure 4.4 Divergence of the PnO method during a rapidly changing irradiance level[37]

Beginning from point A, if the weather conditions remain the same, a change ΔV in the PV voltage will move the operation point to point B, and the direction of the next perturbation of the voltage is reversed due to the reduction in PV power but if the irradiation increases and move the power curve from P1 to P2 the operating point will be relocated from A to C [55]. Point C will be a better power and the PnO will continue to perturb but point C will not be the MPP thus failing to track the MPP. The PnO algorithm is in the family of the hill climbing. It uses the gradient to find the optimum point. This algorithm is not reliable under partial shading conditions as it might not find the GMPP on the P-V curve. As it is widely used in industry this method will be used to show its drawbacks under all atmospheric conditions and partial shading.

4.2.2   Incremental Conductance

Incremental conductance (IC) is a method were by the change of direction of the terminal voltage for the PV panel is determined by comparing the incremental conductance and instantaneous conductance of the PV panel, [56], [57]. The MPP of the PV panel is obtained when the incremental conductance and the instantaneous conductance are equal.

The gradient on the P-V curve of a PV cell (dP/dV) is zero at the MPP, positive on the left of the MPP, and negative on the right as shown in Figure 4.5 [58].

IC uses $\left(\frac{dP}{dV} = 0 \; at \; MPP\right)$ which can be broken down as:

$$\frac{dP}{dV} = \frac{d(VI)}{dV} = I\frac{dV}{dV} + V\frac{dI}{dV} = I + V\frac{dI}{dV} = 0 \tag{4.2}$$

Equation (4.2) can be rearranged to:

$$\frac{dI}{dV} = -\frac{I}{V} \tag{4.3}$$

The equations used for the algorithm include

$$\frac{dI}{dV} = -\frac{I}{V} \qquad \text{at MPP}$$

$$\frac{dI}{dV} > -\frac{I}{V} \qquad \text{left of MPP} \qquad\qquad\qquad (4.4)$$

$$\frac{dI}{dV} < -\frac{I}{V} \qquad \text{right of MPP}$$

The IC flow chart (Figure 4.6) shows how the MPP is tracked.

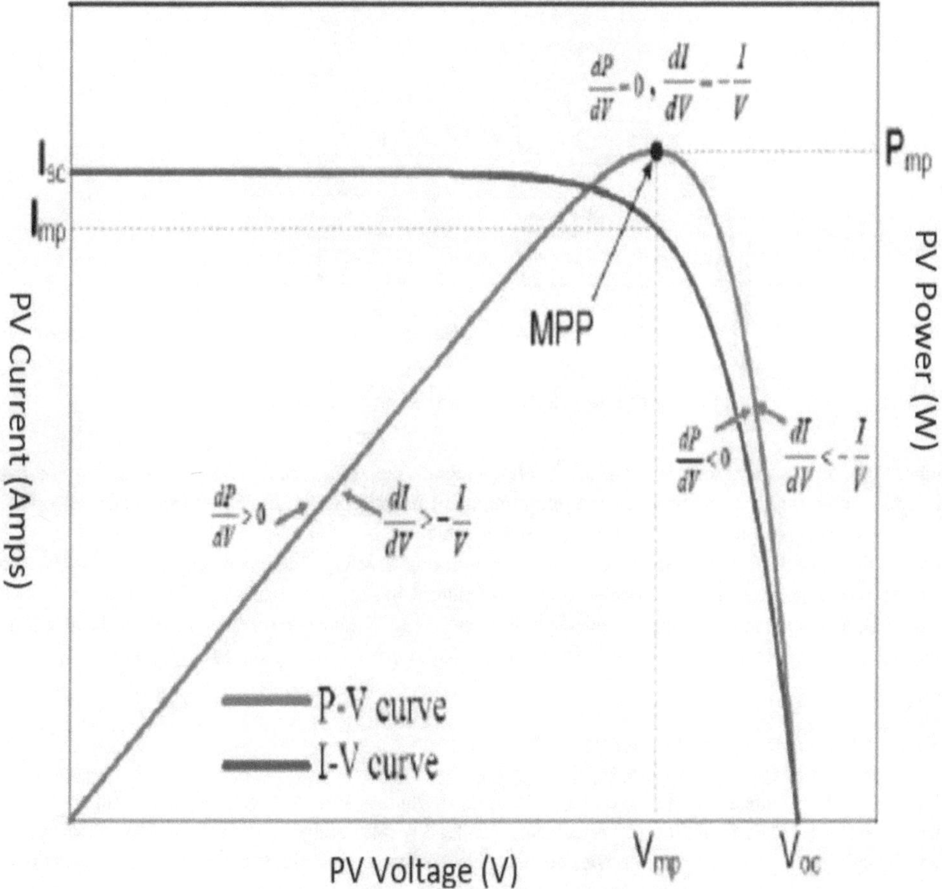

Figure 4.5 :Incremental conductance method on the P-V curve of a solar module [58]

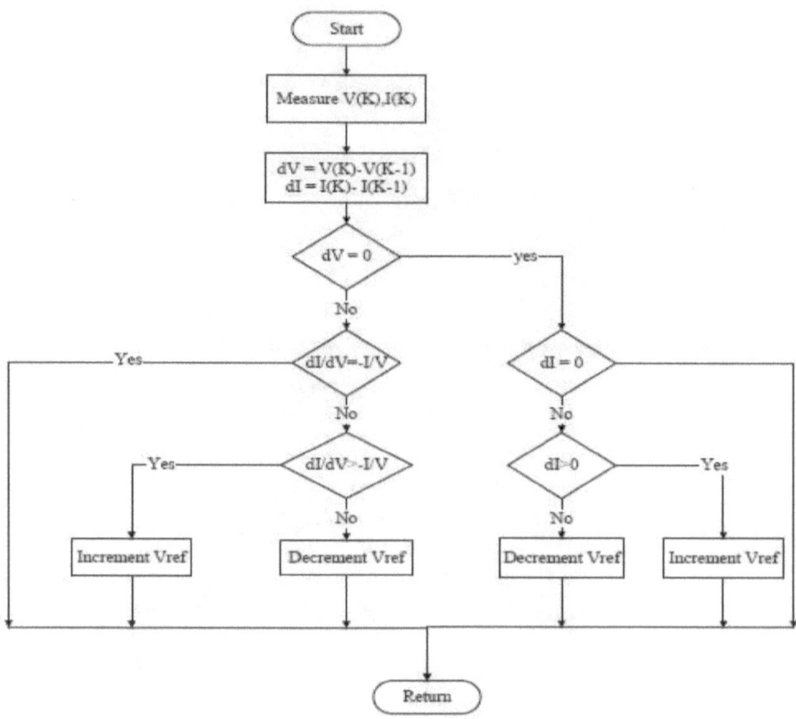

Figure 4.6:The flow chart of the incremental conductance[58]

where Vref equals to maximum power point voltage found Vmpp [53]. Equations (4.4) are used to find the direction of voltage increment when the operating point moves towards the MPP. The increment of the terminal voltage will take place until equation (4.3) is achieved.

When the voltage and current of PV panel change during a voltage increment and $dI/dV > -I/V$, the operating voltage of PV panel is located on the left side of the MPP in Figure 4.5. The operating voltage has to be increased in order to track the MPP. If $dI/dV < -I/V$, the operating voltage of PV panel will be located on the right side of the MPP and has to be decreased in order to find the MPP [57], [59].

4.2.3   Load matching with the PV array

This is one of the simplest methods to make a PV array to operate at MPP. The optimum operating point of the PV arrays is found by theoretical calculation. The load is calculated based on the values of PV voltage and PV current at which maximum power point occurs. The advantage of this method is that no additional circuitry is needed and it is very simple. The challenge of this method is that if one considers the changes of solar irradiation or temperature, it is not reliable [2].

4.2.4   Constant voltage technique (CV)

Ref [60] , [61] applied the fixed voltage technique. A constant voltage reference (Vref) is compared with the PV terminal voltage and the resultant signal is passed through a PI controller. The control output signal is then compared with a triangular waveform to create a duty cycle that controls the converter. The Vref voltage is set to be equal to the V_{MPP} of the array of the characteristic PV array or to another calculated optimum voltage. The drawback of this method is that it does not correct Vref according to the atmospheric variation since the Vref is always fixed [62]. Figure 4.7 shows the schematic diagram.

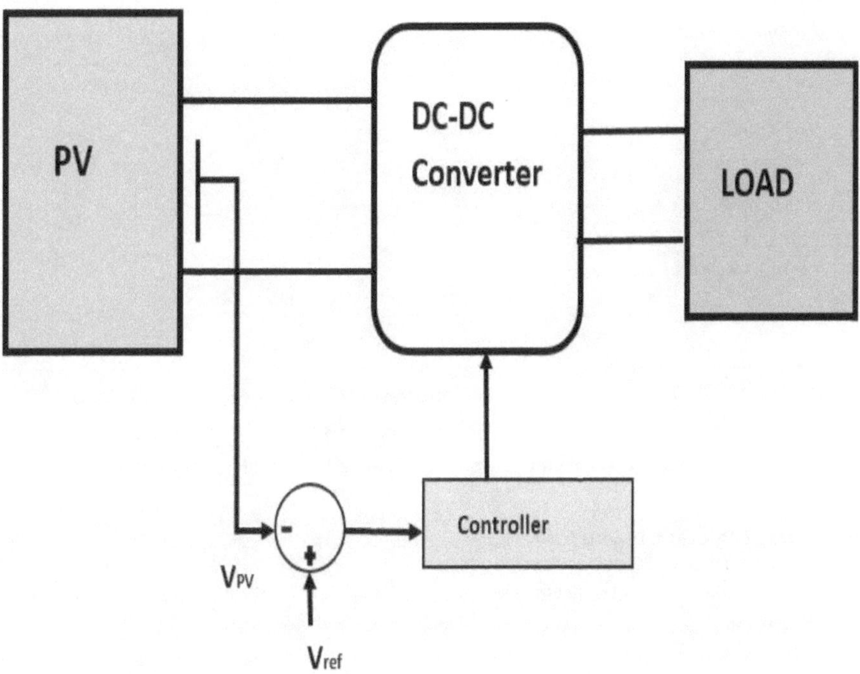

Figure 4.7: Voltage control method for MPPT with fixed voltage reference[60]

4.2.5   Fractional short-circuit current (SC)

The principle of short circuit current technique is based on that under changing weather conditions I_{MPP} can be estimated to be proportional to I_{SC} equation (4.5) illustrates this.

$$I_{MPP} \approx K_{SC}I_{SC} \qquad (4.5)$$

where K_{SC} is a proportional constant. The algorithm flow chart is shown in Figure 4.8 [63]. Once K_{SC} is known and I_{SC} is measured, the I_{MPP} can be calculated from equation (4.5). The advantage of this method is that it's simple and has a low cost of implantation. The drawbacks are that, the maximum power point is never matched [64]

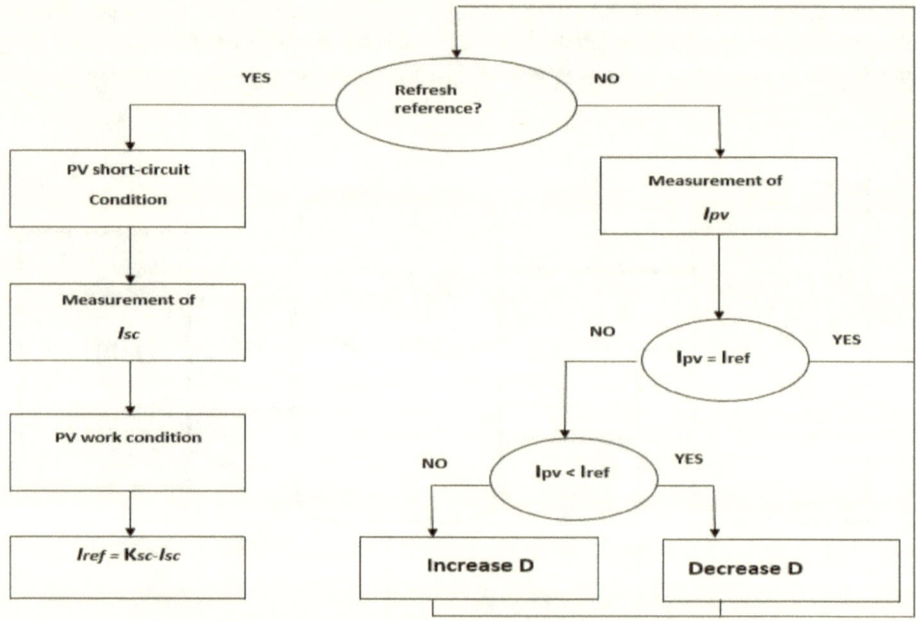

Figure 4.8: Flowchart for Fractional short-circuit current algorithm[64]

4.2.6   Fuzzy Logic MPPT controller

The fuzzy logic has been applied to MPPT with great success as demonstrated in [65], [66], [67]. Fuzzy logic deals with reasoning logic that is approximate rather than fixed and exact. Traditional logic usually sets two–value logic as true or false, but fuzzy logic can have many varying values. Fuzzy logic variables can have a truth or false value that ranges in different degrees and be expressed by linguistic variable [68]. Considering this, fuzzy logic can provide both speed and accuracy [65]. FL method is one of the most efficient techniques in the extraction of the MPP in solar PV systems since it has several advantages [68]. The behaviour of the FLC depends on the shape of Membership Functions (MFs) and the rule-base. Membership functions can have many different shapes. The exact type depends on the actual applications. Figure 4.9 shows the different membership functions. A trapezoidal or triangular waveform should be used for systems that require significant dynamic changes in a short duration of time. The Gaussian waveform can be used for systems that require high control accuracy. In this book, triangular membership functions are used to reduce the computation complexity. The FL controller flow diagram is shown in Figure 4.10. It consists of two inputs and one output. Fuzzy logic controller input variables are the error (E) and change of error (ΔE) and one output variable, change in duty cycle ΔD. The two inputs variables E and ΔE at a sampling point k are given by equation (4.6) and (4.7).

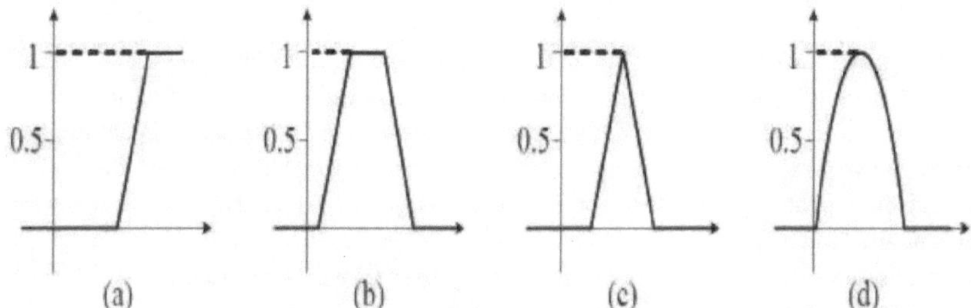

Figure 4.9: Various shapes of membership functions (a) monotonic (b) trapezoidal (c) triangular (d) Gaussian[65]

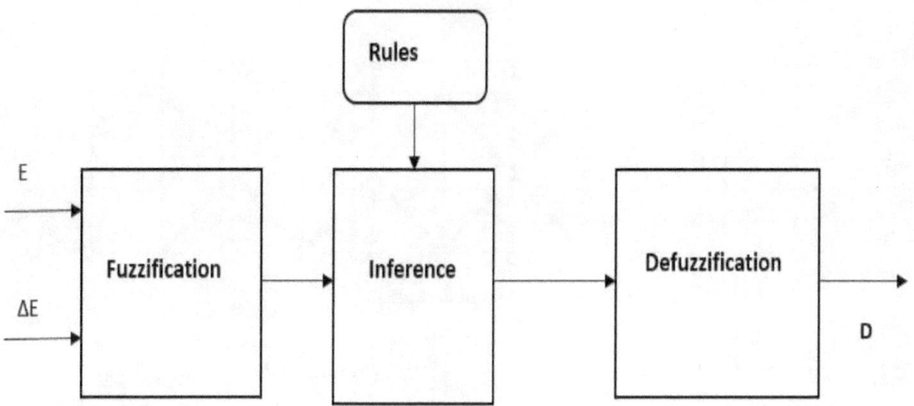

Figure 4.10 The Fuzzy controller diagram[65]

$$E(k) = \frac{P(k)-P(k-1)}{V(k)-V(k-1)} = \frac{\Delta P}{\Delta V} \qquad (4.6)$$

$$\Delta E(k) = E(k) - E(k-1) \qquad (4.7)$$

where *P (k)* is the instant power of the solar PV, and *V (k)* is the voltage at instant *k*. These inputs are chosen so that the instant value of *E(k)* shows whether the load operation power point is located on the right or on the left of the MPP.

If E(k) is positive it means that the operating point is on left side of the MPP and when it is negative, the operating point is on right side of the MPP. The MPP will be obtained when E(k) is equal to zero. While ΔE*(k)* expresses the moving direction of this operation point.
The output of the controller is given by equation (4.8):

$$\Delta D(k) = D(k) - D(k-1) \qquad (4.8)$$

4.2.6.1 Fuzzification

The procedure of converting the system input values E and ΔE into linguistic fuzzy sets using fuzzy membership functions is called fuzzification. The variables are shown in terms of linguistic variables such as ZE (zero), NB (negative big), PB (positive big), NS (negative small) and PS (positive small) using basic fuzzy sub sets as shown in Figure 4.11. Higher number of linguistic variables improves the output stability accuracy but the degree of algorithm complexity is increased. The FLC decides the next operating point depending on the used membership functions and a rule Table 4.2. There are 25 fuzzy control rules used in this book as listed in Table 4.2. IF-THEN statements are used to express the rules.

IF (E is PB) AND (ΔE is NB) THEN (ΔD is NB).

"If E is NB and ΔE is ZE then crisp ΔD is PB it means that if the operating point is far away from the MPP on the right side, and the variation of the slope of the curve is almost Zero; then increase the duty cycle"

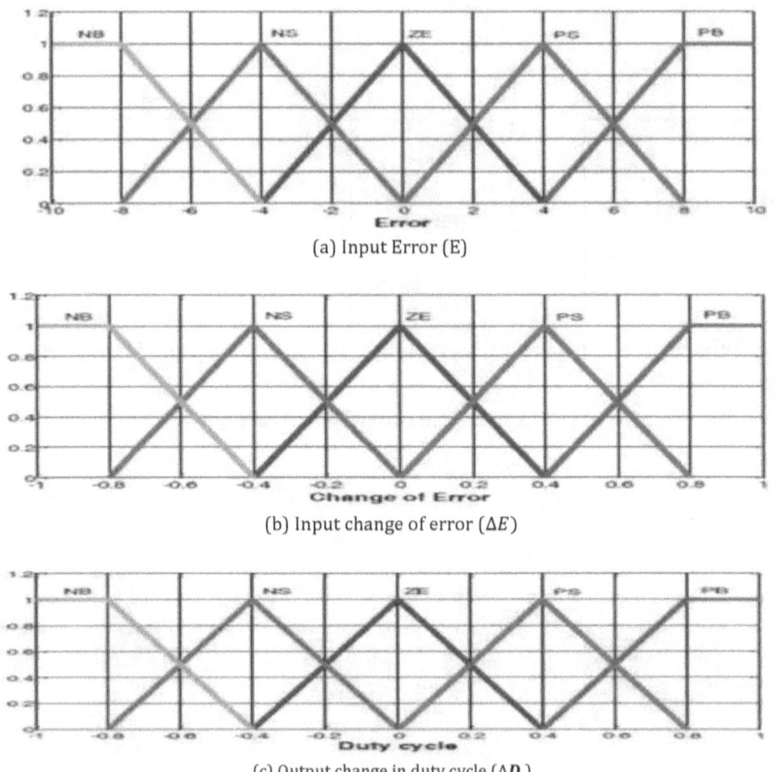

(a) Input Error (E)

(b) Input change of error (ΔE)

(c) Output change in duty cycle (ΔD)

Figure 4.11 Membership functions of E, ΔE and D[68]

"If the value of E(k) is greater than zero, the controller changes the duty cycle to rise the voltage until the power is maximum or the value (ΔP/ΔV) = 0, if the value of E(k) less than zero the controller changes the duty cycle to decrease the voltage until the power is maximum" [65], [68] .

Table 4. 2. Fuzzy logic control rule table[68]

E \ ΔE	NB	NS	ZE	PS	PB
NB	ZE	ZE	PB	PB	PB
NS	ZE	ZE	PS	PS	PS
ZE	PS	ZE	ZE	ZE	NS
PS	NS	NS	NS	ZE	ZE
PB	NB	NB	NB	ZE	ZE

4.2.6.2   Defuzzification

Defuzzification is needed for a crisp output value of control. The defuzzification is implemented using the centroid method (centre of gravity) which is generally employed in the design of fuzzy logics [68]. The basic principle in centre of gravity (CoG) method is to find the point X^*. Figure 4.12 shows the graphical explanation.

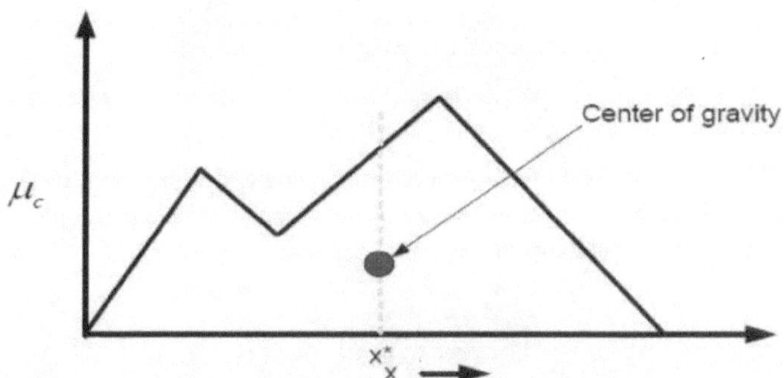

Figure 4.12 Center of gravity method example [68]

Mathematically, the CoG can be expressed as shown in equation (4.9).

$$X^* = \frac{\int X\mu_C(X)dx}{\int \mu_C(X)dx} \qquad (4.9)$$

where X^* is the x-coordinate of centre of gravity and $\int \mu_C(X)dx$ denotes the area of the region bounded by the curve μ_C. If μ_C is defined with a discrete membership function, then CoG can be stated as shown in equation (4.10).

$$X^* = \frac{\sum_{i=1}^{n} X_i \cdot \mu_C(X_i)}{\sum_{i=1}^{n} \mu_C(X_i)} \qquad (4.10)$$

where X_i is a sample element and n is represents the number of samples in fuzzy set C. A geometrical method of calculation [68] can be achieved following the steps bellow.

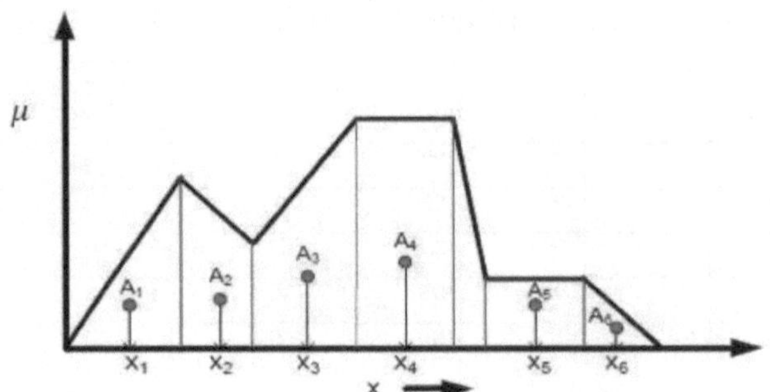

Figure 4.13 Center of gravity method example for discrete membership function [68]

The entire region of Figure 4.13 is divided into a number of small regular regions (e.g. triangles, trapezoid etc.). A_i and X_i denotes the area of the i-th portion. Then X^* according to CoG is calculated as shown is equation (4.11).

$$X^* = \frac{\sum_{i=1}^{n} X_i.(A_i)}{\sum_{i=1}^{n} A_i} \tag{4.11}$$

where n is the number of smaller geometrical components

4.2.7   Unique GMPP

The algorithms discussed so far were for a unimodal P-V curve and they cannot track the GMPP in a multimodal P-V curve. Reference [69] introduced the unique global maximum power point tracker used to track the GMPP under partial shaded conditions.

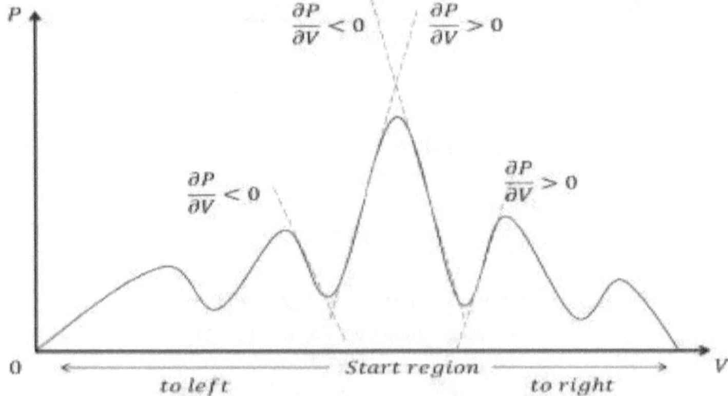

Figure 4.14: Unique global power maximum strategy[69]

Figure 4.14 shows the strategy used to find the GMPP. The algorithm starts performing around the previously stored MPP which was found under uniform atmospheric conditions. It then starts searching from left to right on the curve. During this process it may find several LMPP, if the slopes polarity changes from positive to negative, it means the existence of a LMPP on the left side. In contrast, if the polarity goes from negative to positive, it indicates that an existence of a maximum on the right side. After a local maximum power point is found, it will be compared with the previously stored MPP If the newly found point is better than the previously stored, the new point will be updated to the GMPP and is stored in memory. The major drawback of the unique global power maximum algorithm is the possibility of missing actual global maximum power point when the operating point tracks from current-stored global MPP to the actual global MPP, if there is a small local MPP in between, the tracking process will miss the actual MPP.

4.2.8 Metaheuristic algorithms in MPPT

Metaheuristic algorithms have been applied to MPPT and have shown to perform very good in all atmospheric conditions and partial shaded conditions compared to the conventional algorithms. There major advantage is that they have the ability to search globally, this allows them to find the GMPP on a multimodal P-V characteristic curve of power and also they are population based allowing multiple agents to search for optimality this enables the algorithm to find the GMPP quickly. Examples of these algorithms used in MPPT include, Cuckoo Search (CS) [70], this has been considered one of the best optimisation algorithms recently proposed. In [70], the algorithm was shown to track the GMPP under partial shaded conditions with high accuracy and stability. Reference [41] introduced the Grey Wolf (GW) optimisation to improve lower tracking efficiency, steady state error at MPP experienced by conventional MPPT algorithms. The algorithm was shown to perform well under partial shaded conditions. Other metaheuristic algorithms applied in MPPT include the chaotic search, Ant and Bee Colony (ABC) [17], deferential evolution (DE) [71], Genetic algorithm (GA) [72]. The focus of this book is to use the Particle swarm optimisation and the Firefly algorithm to track the GMPP under partial shaded conditions, these are well known metaheuristic algorithms that have been proven to track the GMPP in PV systems. Emphasis is put on tracking the GMPP, reducing the steady state error and increasing speed and accuracy.

4.3 Control of the Boost Converter

In this section the study of the dynamic modelling of the boost converter is explained using the state space average model and the simulation results are done in chapter 6. The transfer function of the boost converter equation (2.16) is derived from the steady state condition. The steady state occurs after the converter has run for a considerable time then it settles down to a stable condition. The steady-state solution is used to design the converter, as it is the normal condition of operation [26] but this would be an open loop design. The closed loop design also requires a dynamic model of converter.

A converter may be used as a voltage regulator or current controller by closing a feedback loop between the required output quality (output voltage, output current or input current) and switching device's duty-ratio signal (as the control signal). The feedback is compared with a reference for the control. A typical voltage mode regulator is shown below in Figure 4.15.

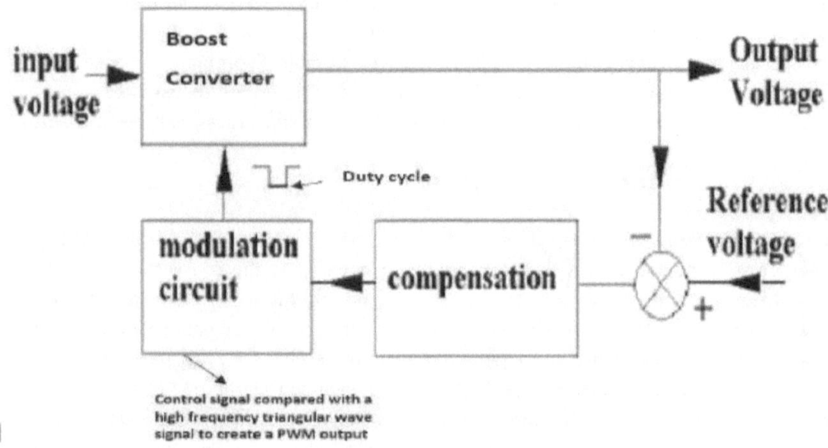

Figure 4.15 Voltage control of a DC-DC converter[26]

The knowledge of the dynamic characteristics of the converter is important for the stable operation of the controller design. This is the challenge of switched-mode converter because the converter is highly nonlinear [26]. An approximate representation of the converter can be obtained using the state- space averaging technique. This technique allows us to obtain a small signal linear system that can be used for linear controller design.

4.3.1 State space averaging technique

There are only two states in the continuous mode of operation i.e. switch is either on or off. The circuit averaging method can be used to find the average of the large signal of the system of the combined states. When the switching device is turned on, it conducts for a ratio D of a period. The state space equation can be written as shown in equation (4.12) and (4.13). These equations are the large signal models in state space meaning the actual signal.

$$\dot{X} = A_{on}X + B_{on}U \tag{4.12}$$

$$Y = C_{on}X \tag{4.13}$$

where X is the state space variables such as capacitor voltage and inductor current. A_{on}, B_{on} and C_{on} are the state space matrices parameter characteristics of the converter during the on states respectively. U is the input variable such as the input voltage V_{in}. Y is the output such as voltage V_o.

When the switch device is turned off, the diode conducts for a ratio of (1-D) of a period. The state space equation can be written as equations (4.14) and (4.15) [71].

$$\dot{X} = A_{off}X + B_{off}U \tag{4.14}$$

$$Y = C_{off}X \tag{4.15}$$

where X, U, and Y are the state space, input and output variables as defined previously and A_{off} and B_{off} and C_{off} are the state -space matrices parameter characteristics of the converter during the off states respectively. The output matric C_{off} is the same for both states.

The two states of ON and OFF represented by D and 1-D duration can be averaged by the conduction ratio shown in equation (4.16) to form the average large signal model.

$$\dot{X} = [DA_{on} + (1-D)A_{off}]X + [DB_{on} + (1-D)B_{off}]U \qquad (4.16)$$

If there is a small signal d applied to the steady state duty ratio $\bar{D}$ this will cause a small variation x of the steady state variable $\bar{X}$. The following equations can be assumed for the large signal (actual signal).

$$D = \bar{D} + d \qquad (4.17)$$

$$X = \bar{X} + x \qquad (4.18)$$

$$U = \bar{U} + u \qquad (4.19)$$

where x is a small variation on the steady sate component $\bar{X}$ and d is a small signal variation in the steady state component $\bar{D}$, u is the small signal input and $\bar{U}$ is the steady state input. Substitution of these parameters in to equation (4.16) gives off equation (4.20) [73].

$$\dot{\bar{X}} + \dot{x} = [(\bar{D} + d)A_{on} + (1 - \bar{D} - d)A_{off}][\bar{X} + x] + [(\bar{D} + d)B_{on} + (1 - \bar{D} - d)B_{off}](\bar{U} + u) \quad (4.20)$$
This can be expanded to

$$\dot{\bar{X}} + \dot{x} = [\bar{D}A_{on} + (1 - \bar{D})A_{off}]\bar{X} + [\bar{D}B_{on} + (1 - \bar{D})B_{off}]\bar{U} + u$$
$$+ [dA_{on} - dA_{off}]\bar{X} + [dB_{on} - dB_{off}]\bar{U} + u$$
$$+ [\bar{D}A_{on} + (1 - \bar{D})A_{off}]x + [dB_{on} - dB_{off}]x \qquad (4.21)$$

Equation (4.16) can be written as follows if the small signal is approximated to be close to zero.

$$\dot{\bar{X}} = [\bar{D}A_{on} + (1 - \bar{D})A_{off}]\bar{X} + [\bar{D}B_{on} + (1 - \bar{D})B_{on}]\bar{U} + u \qquad (4.22)$$

Removing equation (4.22) from equation (4.21) and neglecting the high order small signal variation, this will result in equation (4.23) the small signal model which is linear and can be used for controller design. This is the one that gives us the dynamics of the system i.e variations about the operating point [26], [73].

$$\dot{x} = Ax + Bu + Fd \qquad (4.23)$$
where A is the average of A_{on} and A_{off} and F is the difference of equations (4.12) and (4.14). Equation (4.23) is the linearized equation.
 where d forms the control input.
The Small signal output becomes
$$y = Cx \qquad (4.24)$$
where

$$B = \bar{D}B_{on} + (1 - \bar{D})B_{off} \tag{4.25}$$

$$A = \bar{D}A_{on} + (1 - \bar{D})A_{off} \tag{4.26}$$

$$F = [A_{on} - A_{off}]X + [B_{on} - B_{off}]U \tag{4.27}$$

The equation can be solved by conventional Laplace transform [74].

$$\frac{x}{d} = [sI - A]^{-1}F \tag{4.28}$$

where I the unit matrix. $[sI - A]^{-1}$ is the inverse of $[sI - A]$

Using the above equations, the small signal transfer function of the boost converter can be obtained.

4.3.1.1 Small signal of the boost converter

The parameters of the energy storage elements are used for the state space variable. Using Figure 2.22 and applying Kirchhoff's law, the following state equations can be obtained [26].

When T is on as seen in Figure 2.22 the state equations (4.29) and (4.30) are obtained.

$$V_{in}(t) = L\frac{dI_L(t)}{dt} \tag{4.29}$$

$$\frac{-V_c}{R} = C\frac{dV_c(t)}{dt} \tag{4.30}$$

Rearranging the equation and making the state variables the subject of the formula. This can be expressed in state space form as shown in equation (4.31)

$$\begin{bmatrix} \dot{i}_L \\ \dot{v}_c \end{bmatrix} = \begin{bmatrix} 0 & 0 \\ 0 & \frac{-1}{RC} \end{bmatrix}\begin{bmatrix} i_L \\ v_c \end{bmatrix} + \begin{bmatrix} \frac{1}{L} \\ 0 \end{bmatrix}V_{in} \qquad \equiv X = A_{on}X + B_{on}U \tag{4.31}$$

When T is off, the state equations become

$$\frac{dI_L(t)}{dt} = \frac{V_{in}}{L} - \frac{V_c}{L} \tag{4.32}$$

$$\frac{dV_c}{dt} = \frac{-V_c}{CR} + \frac{I_L}{C} \tag{4.33}$$

$$\begin{bmatrix} \dot{i_L} \\ \dot{v_c} \end{bmatrix} = \begin{bmatrix} 0 & \dfrac{-1}{L} \\ \dfrac{1}{C} & \dfrac{-1}{RC} \end{bmatrix} \begin{bmatrix} i_L \\ v_c \end{bmatrix} + \begin{bmatrix} \dfrac{1}{L} \\ 0 \end{bmatrix} V_{in} \qquad \equiv \qquad X = A_{off} X + B_{off} U \tag{4.34}$$

For the output equation voltage control is selected but current control is also an option [73]. The output is the same in both states that is

$$V_o = [0 \ 1] \begin{bmatrix} i_L \\ V_c \end{bmatrix} \tag{4.35}$$

The averaged large signal model becomes the combination of the two states and obtaining the equivalent values. Substitute the state equations of the converter into equation (4.16) gives

$$\begin{bmatrix} \dot{i_L} \\ \dot{v_c} \end{bmatrix} = \begin{bmatrix} 0 & \dfrac{-(1-D)}{L} \\ \dfrac{1-D}{C} & \dfrac{-1}{RC} \end{bmatrix} \begin{bmatrix} i_L \\ V_c \end{bmatrix} + \begin{bmatrix} \dfrac{1}{L} \\ 0 \end{bmatrix} [V_{in}] \tag{4.36}$$

The output is seen in equation (4.35). Linearization takes place when we remove the steady state component from the large signal model and neglect the high order deviations to obtain the linear small signal model.

In the steady state condition ($\dot{\overline{X}} = 0$) is equal to zero.

From equation (4.36) the steady state model can be obtained. Note the derivative terms are equated to zero and only the steady state variable of inductor current and capacitor voltage are considered.

$$\begin{bmatrix} 0 \\ 0 \end{bmatrix} = \begin{bmatrix} 0 & \dfrac{-(1-\overline{D})}{L} \\ \dfrac{1-\overline{D}}{C} & \dfrac{-1}{RC} \end{bmatrix} \begin{bmatrix} \overline{i_L} \\ \overline{v_c} \end{bmatrix} + \begin{bmatrix} \dfrac{1}{L} \\ 0 \end{bmatrix} [V_{in}] \tag{4.37}$$

To get the transfer function relationship consider

$$0 = A\overline{X} + B\overline{U} \quad = \overline{X} = -A^{-1} B\overline{U} \tag{4.38}$$

$$Y = C\overline{X} = -CA^{-1} B\overline{U} \tag{4.39}$$

So Y/U will give us the output input relationship.

Equation (4.35) will be equal to equation (4.39)

$$V_o = CA^{-1} \begin{bmatrix} \dfrac{1}{L} \\ 0 \end{bmatrix} [V_{in}] \tag{4.40}$$

So for the steady state input output relationship using equation (4.37)

$$V_0 = -[0\ 1] \begin{bmatrix} 0 & 0 \\ \frac{-L}{1-\bar{D}} & 0 \end{bmatrix} \begin{bmatrix} \frac{1}{L} \\ 0 \end{bmatrix} [V_{in}] = \frac{-L}{(1-\bar{D})} \times \frac{1}{L} \qquad (4.41)$$

$$\frac{\bar{V_0}}{\bar{V_{in}}} = \frac{1}{(1-\bar{D})} \qquad \qquad used\ for\ Boost\ Design \qquad (4.42)$$

So for the small signal transfer function using equation (4.36) we substitute the small signal deviations. Equation (4.43) shows the steady state component with small signal component

$$\begin{bmatrix} \dot{\bar{\iota}_L + \iota_L} \\ \bar{v}_c + v_c \end{bmatrix} = \begin{bmatrix} 0 & \frac{-1+\bar{D}+d}{L} \\ \frac{1-\bar{D}-d}{C} & \frac{-1}{RC} \end{bmatrix} \begin{bmatrix} \bar{\iota}_L + i_L \\ \bar{v}_c + v_c \end{bmatrix} + \begin{bmatrix} \frac{1}{L} \\ 0 \end{bmatrix} [\bar{V}_{in} + v_{in}] \qquad (4.43)$$

$$\bar{V_0} + v_0 = [0\ 1] \begin{bmatrix} \bar{\iota}_L + i_L \\ \bar{v}_c + v_c \end{bmatrix} \qquad (4.44)$$

So if we remove the steady state part we left with the small signal model shown in equation (4.45) and (4.46)

$$\begin{bmatrix} \dot{\iota_L} \\ v_c \end{bmatrix} = \begin{bmatrix} 0 & \frac{-1+\bar{D}}{L} \\ \frac{1-\bar{D}}{C} & \frac{-1}{RC} \end{bmatrix} \begin{bmatrix} i_L \\ v_c \end{bmatrix} + \begin{bmatrix} \frac{1}{L} \\ 0 \end{bmatrix} [v_{in}] + \begin{bmatrix} \frac{v_c}{L} \\ \frac{-i_L}{C} \end{bmatrix} d \qquad (4.45)$$

$$v_0 = [0\ 1] \begin{bmatrix} i_L \\ v_c \end{bmatrix} \qquad (4.46)$$

where
i_L, v_c and d are small-signal perturbations about the steady-state values of $\bar{\iota}_L, \bar{v}_c$ and $\bar{D}$ respectively.

The transfer function we require would be controlling the output voltage using the control input of duty cycle. This is known as voltage control of boost converters. Using Laplace transforms the following equation can be used to get the transfer function [74].

$$Y(s) = C(sI - A)^{-1} BU(s) \qquad (4.47)$$

So for the transfer function of output voltage and duty cycle input used for voltage control.

$$\frac{v_o}{d}(s) = C(sI - A)^{-1} B \qquad used\ for\ controller\ design\ (Voltage\ control) \qquad (4.48)$$

where B in the input matrix for the duty cycle input (d)

Matlab m file (appendix C) was used to get the transfer function of the designed boost converter from the state space model.

4.3.2 Controller design

The controller design is based on the fact that the small signal model is linear. The majority of industry control is still done by Proportional integral (PI) or Proportional integral derivative (PID). These linear controllers are still able to do the job despite other numerous advancements in control systems. Their ability to work over a wide range of operating conditions makes them favourable [74]. The parameters of these controllers are well understood. The stability of the system will depend on the controller. Table 4.3 shows the purpose of each parameter [82]. The parameters obtained of PI are for the small signal model not the large signal model (the actual system) which is nonlinear.

Table 4. 3:Effects of controller parameters[74]

Close Loop Response	Rise Time	Overshoot	Settling Time	Steady state error
P	decrease	increase	Small change	decrease
I	decrease	increase	increases	eliminate
D	Small change	decrease	decrease	Small change

The meaningful parameters used to define the behaviour of a system response are: [74]

- Rise Time ; the time taken for the output voltage to go from 10% to 90 % of the final value
- Peak Time; the time taken for the output to reach its maximum value.
- Percentage overshoot; (max value- final value)/final value $\times$ 100.
- Settling time
- Steady state error; the difference between the set point reference and the final value achieved.

A typical response of a second order system is shown in Figure 4.16. It also shows the defined parameters. Typically when designing a control system, you have to achieve certain targets for these parameters [74].

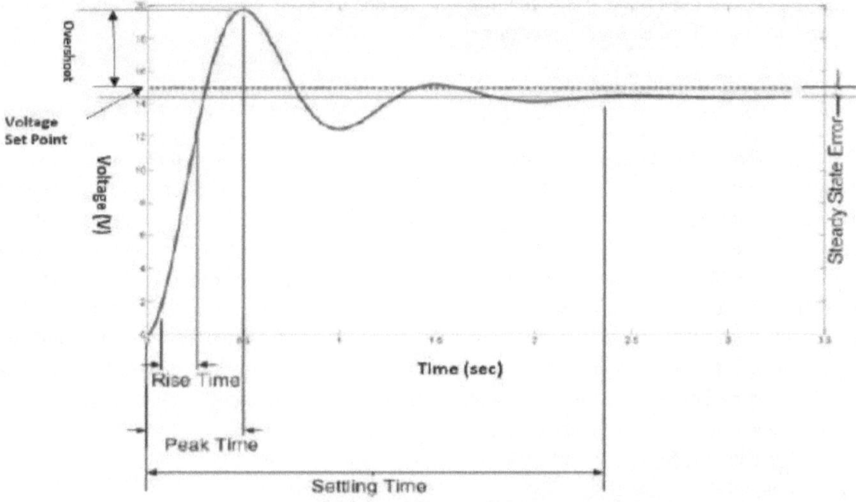

Figure 4.16: A typical response to a second order system[74]

Figure 4.17 shows how the complete system looks for a boost converter where the output voltage is being controlled by the duty cycle signal (d). Vref is the desired reference point.

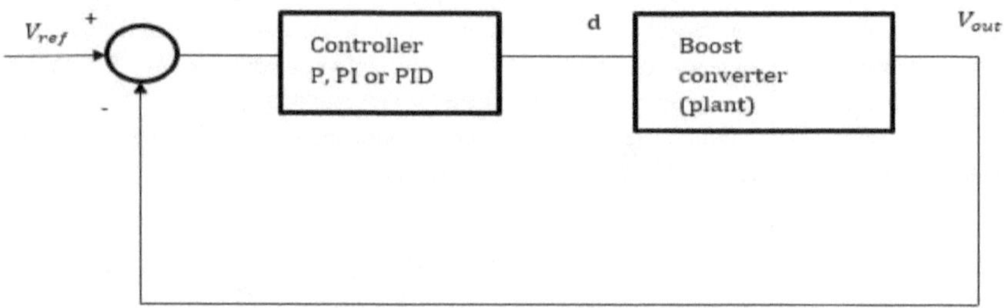

Figure 4.17: Controller setup[74]

The PID controller can be represented as shown below in its Laplace transform state

$$G(s) = K_p + \frac{K_i}{s} + K_d s \qquad (4.46)$$

K_p is the proportional gain ,K_i is the integral gain and K_d is the derivative gain. Proportional control is defined as the control action that occurs in direct proportion with the system error. A large proportional gain will result in a small error and hence a large control signal.

The integral action allows the achievement of equality between the measured value and the desired value .The derivative action means that changes in the desired value may be anticipated [74].In most plants the derivative gain is not necessary and the controller just becomes a PI controller. Classical methods such as Ziegler Nichols can be used to find the optimum values of the gain parameters the performance of the closed loop system will be highly dependent on these gain parameters also optimisation algorithms like the metaheuristics can be used to find these optimum gain parameters but using these algorithms for optimum PI controller tuning will not be the focus of this book.

4.4 Summary

This chapter reviews and discuss the need for MPPT in PV systems. Classical MPPT methods as well as computational intelligence methods are discussed. The comparison suggests that, computational intelligence methods have the advantage of searching for the GMPP under partial shading conditions. It is also found that under rapid increasing or decreasing of irradiance the PnO can get lost in tracking the MPP. Computational intelligence algorithms have been successfully been implemented in MPPT showing improved results. At steady state they converge to a particular optimum voltage or duty cycle value resulting in less oscillations compared to the fixed step size in PnO. The dynamic response is also superior because of the multiple agents looking for the MPP. In the literature the tracking efficiency of CI algorithms is also found to be better than the conventional optimisation algorithms. They converge closer to the required solution value resulting in producing a higher output PV power.

The chapter also discuss how the small signal model of the boost converter is derived from the state space average technique.

Chapter 5: Particle Swarm Optimisation and The Firefly algorithm

In this chapter the Firefly optimisation algorithm (FA) and the particle swarm optimisation (PSO) algorithm are introduced. They are then used in the implementation of the MPPT to find the optimum power under all atmospheric conditions. PSO and FA do not require gradient information of the objective function being optimized to find the optimum value [43] as opposed to the conventional algorithms.

5.1 The Firefly algorithm

The Firefly algorithm is a metaheuristic algorithm [29]. This algorithm was inspired by the behavior of flashing fireflies [29].

The light intensity at a distance r from the light source obeys the inverse square law. That is light intensity (I) keeps on decreases as the distance (r) increase (I α $\frac{1}{r^2}$). Additionally, the *"air keeps absorbing the light which becomes weaker with the increase in the distance. These two factors when combined make most fireflies visible at a limited distance, normally to a few hundred meters at night, which is quite enough for fireflies to communicate with each other"* [28], [29].

5.1.1 Concept

The algorithm follows the three basic concepts.

- All the fireflies are unisex [29].
- *"Attractiveness and brightness are proportional to each other, so for any two flashing fireflies, the less bright one will move towards the one which is brighter. Attractiveness and brightness both decrease as their distance increases"* [29]. If the fireflies have the same brightness they will move randomly.
- The brightness of the firefly is based on where it is located in the objective function.

5.1.2 Light intensity and attractiveness

There are two important concepts in the FA these are the variation in the light intensity and how the attractiveness is obtained [29]. To simplify the formulation the attractiveness of a firefly is obtained based on its brightness. The brightness is related to the objective function being optimized [29]. The attractiveness β is relative and should be judged by the other fireflies. Hence, it will change with the distance r_{ij} between firefly i and firefly j. The attractiveness will vary with the degree of absorption of light in the media [29]. Light intensity I(r) changes according to the inverse square law shown in equation (5.1).

$$I(r) = \frac{I_s}{r^2} \tag{5.1}$$

where I_s is the source intensity. Given a medium with a constant light absorption coefficient γ, the light intensity *I* will change with distance *r* as shown in equation (5.2) [29].

$$I = I_0 e^{-\gamma r} \tag{5.2}$$

where I_0 is the original light intensity. In the expression $\frac{I_s}{r^2}$ singularity at r=0 can be avoided by the combined effect of both equation (5.1) and absorption can be estimated by the Gaussian form

$$I(r) = I_0 e^{-\gamma r^2} \tag{5.3}$$

Since the firefly attractiveness varies linearly to the intensity experienced by adjacent fireflies. Attractiveness of a firefly can now be obtained by

$$\beta = \beta_0 e^{-\gamma r^2} \tag{5.4}$$

where β_0 is the attractiveness at r = 0. Equation (5.4) defines a characteristic distance $\Gamma = \frac{1}{\sqrt{\gamma}}$ over which the attractiveness varies significantly from β_0 to $\beta_0 e^{-1}$

The attractiveness function $\beta(r)$ is selected to be a monotonically decreasing function. Equation (5.5) shows the basic form.

$$\beta(r) = \beta_0 e^{-\gamma r^m}, \quad (m \geq 1) \tag{5.5}$$

where m is an integer greater than 1
For a constant γ, the characteristic length is

$$\Gamma = \gamma^{-1/m} \longrightarrow 1, \, m \longrightarrow \infty \tag{5.6}$$

The distance that separates any two fireflies i and j at x_i and x_j, respectively, is the Cartesian distance

$$r_{ij} = \| x_i - x_j \| = \sqrt{\sum_{k=1}^{d}(x_{i,k} - x_{j,k})^2} \tag{5.7}$$

where $x_{i,k}$ is the kth component of the spatial coordinate x_i of ith firefly, d is the dimension space. In 2-D case, we have

$$r_{ij} = \sqrt{(x_i - x_j)^2 + (y_i - y_j)^2} \tag{5.8}$$

Firefly i will be attracted to another more attractive (brighter) firefly and will move towards it according to equation (5.9)

$$x_i = x_i + \beta_0 e^{-\gamma r_{ij}^2} (x_i - x_j) + \alpha \, \epsilon_i \tag{5.9}$$

where the second term in Equation (5.9) represents attraction. The last term in equation (5.9) is randomization with α being the parameter used to randomise, and ϵ_i is a vector of random numbers

[29]. Equation (5.9) represents a random walk biased towards the brighter fireflies. If $\beta_0 = 0$, it becomes a random walk. For most implementation, we can take $\beta_0 = 1$. Figure 5.1 below shows the movement of the firefly.

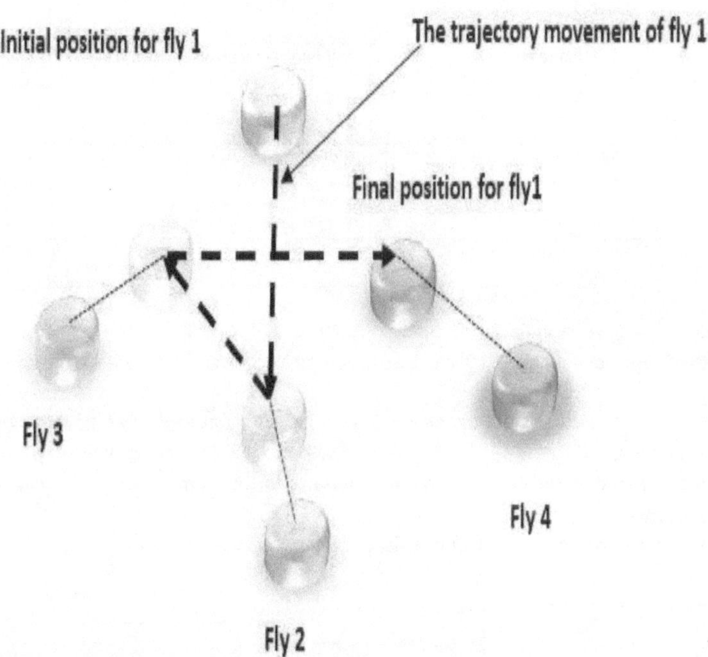

Figure 5.1 Trajectory movements of firefly 1[43]

5.1.3 Algorithm parameters

Proper tuning of the FA parameters is very necessary for global optimum searching [28]. The most important parameter in firefly algorithm is γ, it characterizes the variation of the attractiveness and it plays a very crucial role of how fast the algorithm converges and how the algorithm behaves [43]. In theory γ ∈ [0, ∞] but when implementing γ is obtained by the characteristic length Γ of the system to be optimised. γ usually varies from 0.1 to 10 for most applications. The parameter α allows the search process to escape the local search and search on the global scale [28], [29], [43].

5.1.4 Firefly algorithm in MPPT in PV systems

In implementing the algorithm in MPPT, the PV voltage is chosen as the regulated variable and the objective function is chosen as the PV output power. The table 5.1 below shows how the FA terminologies match those of the PV systems.

FA algorithm	PV system
Firefly position	Voltage reference (V_{ref})
Distance	Voltage difference (ΔV_{ref})
Attractiveness	Exponential function of (ΔV_{ref})
Brightness	Power (P_{pv})
Brightest Firefly	Global maximum power (P_{Gbest})
Dimension(number of variables)	One

The position(x_i) variable in equation (5.9) is considered as the voltage reference (V_{ref}), whereas the second and third terms of equation (5.9) can be considered as correcting the voltage references values.

The purpose of the firefly is to obtain the best V_{ref} voltage value that gives the best PV power. The voltage controller method is used to find the MPP. This is shown in Figure 5.2 along with the whole system. The best V_{ref} value obtained by the FA is used as the reference for the comparator and the varying PV voltage tracks this reference. The error is sent to the PI controller and the output of the control signal from the PI is used to create the pulse with modulation signal to switch the boost switches.

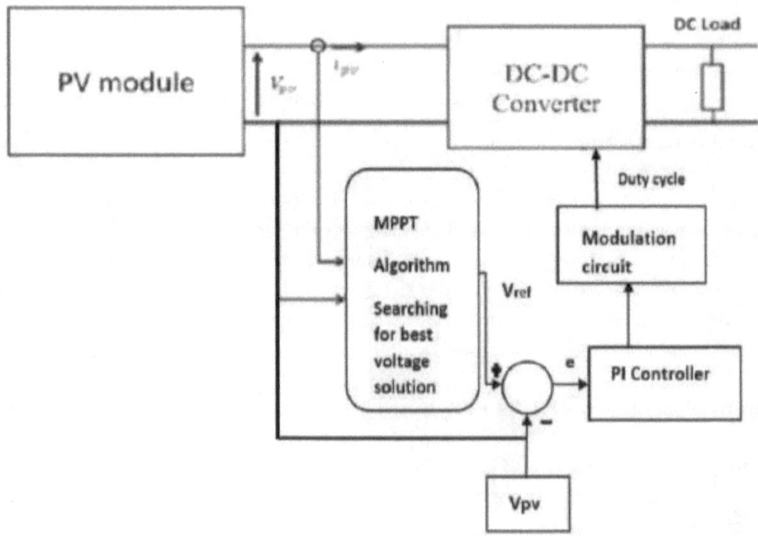

Figure 5.2 The voltage control method[43]

First, V_{ref} is initialized and the number of fireflies is set to N. The power level of each firefly determines how it will be ranked. The global maximum power P_{Gbest} is initialised according to this power ranking. In the inner loop of Figure 5.3, the intensity of fly *i* is compared with that of fly *j* and *j≠ i*.

If Ppv, *j* > Ppv,*i* , the voltage reference is updated by equation (5.9).
For MPPT (dimension) d=1 Hence r_{ij} in equation (5.7) becomes

$$r_{ij} = \sqrt{(x_i - x_j)^2} \tag{5.10}$$

P_{Gbest} is updated by ranking the flies. Equation (5.11) is the GMP convergence criterion.

$$|P_{Gbest} - P_{pv,i}| < \epsilon \ \ i = 1 \dots N \tag{5.11}$$

where ϵ is the tolerance value [43].

The flow chart below shows how the FA is implemented for the MPPT.

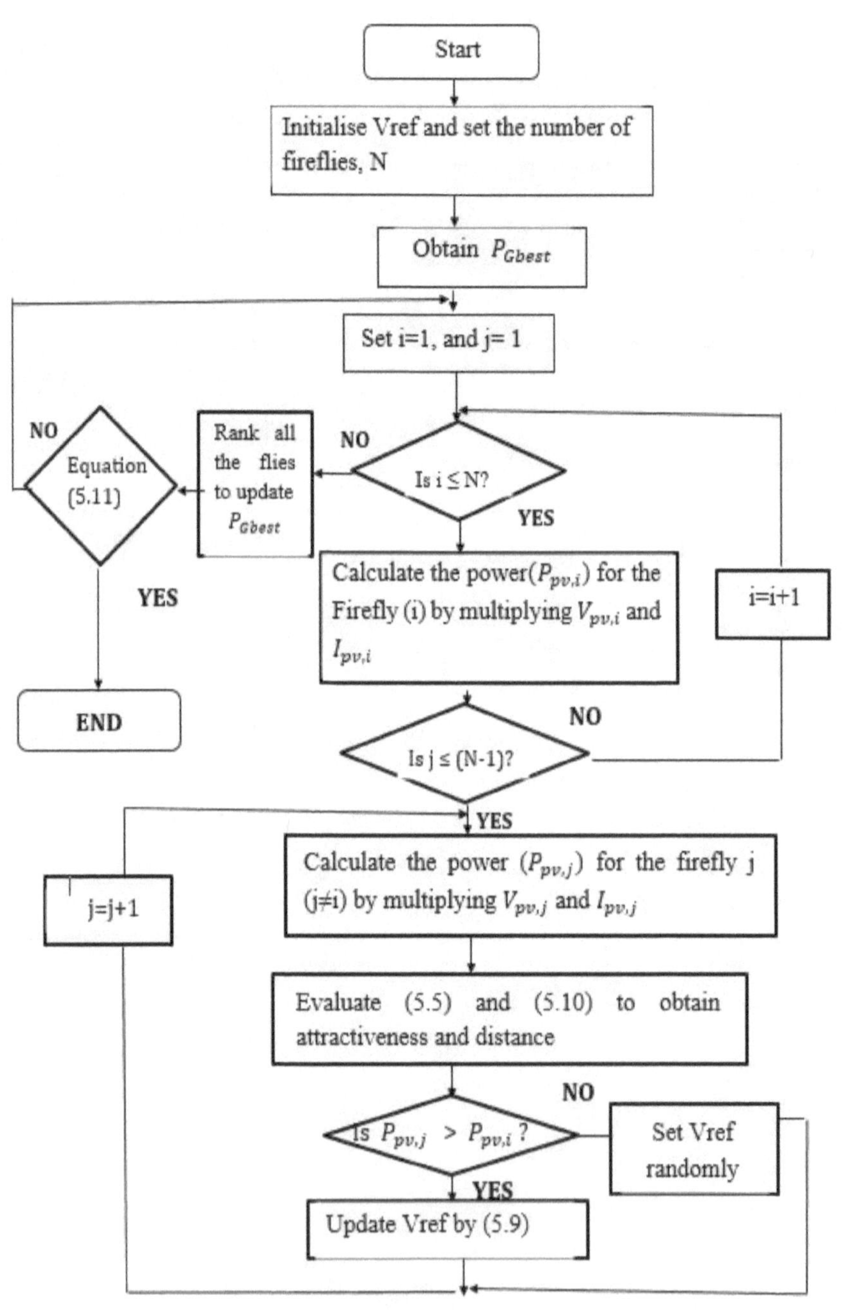

Figure 5.3 Firefly Flow chart for MPPT[43]

5.2   Particle Swarm Optimization

PSO is a stochastic, population-based algorithm based on the movement of fish when they feed in nature. The PSO uses a population of agents (particles), where each agent is a candidate solution [29].

Each particle has to obey two rules these are to follow the best performing particle in the swarm, and to move toward the best conditions found by the particle itself. By this way, each particle eventually evolves to an optimal or close to optimal [29]. The location of a particle is, therefore, determined by the best particle in the neighbourhood *P* best as well as the optimum solution discovered by all the particles in the entire swarm *G* best. The particle position *xi* is modified by using equation 5.12. Figure 5.4 shows the typical movement of particles in the optimization process.

$$x_i^{k+1} = x_i^k + \varphi_i^{k+1} \tag{5.12}$$

The velocity is calculated by equation (5.13)

$$\varphi_i^{k+1} = w\varphi_i^k + c_1 r_1 \{ Pi_{best} - x_i^k \} + c_2 r_2 \{ G_{best} - x_i^k \} \tag{5.13}$$

where (w) is the inertia weight, *r1* and *r2* represent random variables distributed uniformly between [0, 1]; and *c1, c2* are the cognitive and social coefficient, respectively. *Pi* best is the personal optimum location of particle *i*, and *G* best is the optimum location of the particles in the entire population [28], [75].

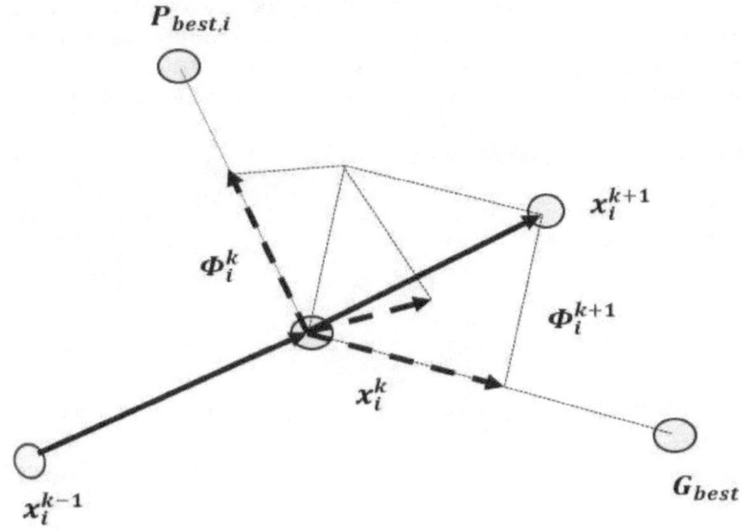

Figure 5.4 :Movement of particles in the optimisation process[75]

5.2.1 Algorithm parameters

The first part of equation (5.13) is used to control the convergence behaviour of the PSO [28], [76]. The second part in equation (5.13) represents the personal ability of each individual particle. The cognitive component makes the particles biased towards their own best positions currently found [76]. The third part is the social component which represents how the swarm population come together to find the global best solution. The social component always attracts the particles toward the global optimum particle found so far [77].

Time varying parameters are introduced i.e., c_1, c_2 and w to effectively control the global search and convergence to the global best solution [77], [78]. Time varying parameters showed a significant performance of the PSO method [79]. A linearly varying inertia weight (w) over the generation is used. The mathematical representation of this concept is given by equation (5.14). The inertia weight is chosen such that the effect of φ_i^k fades during the execution of equation (5.13). Thus, it is preferred to reduce the inertia weight values with time [43]. A common practise is to initially have a large inertia weight value for a better global search process and gradually reduce it to get refined solutions [80].

$$w(k) = w_{max} - (w_{max} - w_{min}) \times \frac{(iter)}{MaxIt} \tag{5.14}$$

where w_{min} and w_{max} are the lower and upper values of the inertia weight, respectively.

The search towards optimum convergence depends on tuning of the cognitive component and the social component. Kennedy and Eberhart [81] illustrated that excessive wandering of particle solutions through the search space will occur if a relatively high value of the cognitive component is selected compared with the social component. The cognitive component is increased and the social component is decreased at the beginning this will allow the particles to explore the search space and not converge towards the particle best [79]. At the latter part of the optimization, having a small $c1$ and a big $c2$ will make the particles to converge to the global best.
This is mathematically represented as follows [75].

$$c_1(k) = c1_{max} - (c1_{max} - c1_{min}) \times \frac{iter}{MaxIt} \tag{5.15}$$

$$c_2(k) = c2_{min} + (c2_{max} - c2_{min}) \times \frac{iter}{MaxIt} \tag{5.16}$$

where $c1_{min}, c1_{max}, c2_{min}$ and $c2_{max}$ are constants. They represent the lower and upper bounds of each parameter.

5.2.2 PSO algorithm for MPPT in PV systems

The voltage controller method is used again shown in Figure 5.2. The particle position x_i is considered as V_{ref}. The velocity component is the correction factor to find optimum value of V_{ref}. The objective function to optimize is output Power.

Figure 5.5 shows the flow chart of how the PSO is used for MPPT

The power of each particle Ppv,i is obtained by multiplying the sampled voltage ($Vpvi$) and current ($Ipvi$) coming from the PV array. The algorithm then checks whether this voltage reference value will produce a better PV power which is its individual fitness value by evaluating equation (5.17) [41], [89]:

$$P_{pvi} > P_{pv}, i - 1 \qquad (5.17)$$

If (5.17) is satisfied, the individual fitness value ($pbesti$) is updated; otherwise, $pbesti$ keeps its current value.
"Ppv,i is then checked against the power of the other particles to see if the global fitness value (gbest) requires updating"[89]. Finally, the convergence criterion is defined in equation (5.18). The equation is used to check that all the particles converge to the GMP.

$$| P_{pv,gbest} - P_{pv,i} | < \epsilon \; i = 1 \dots n \qquad (5.18)$$

where ϵ is the tolerance value.

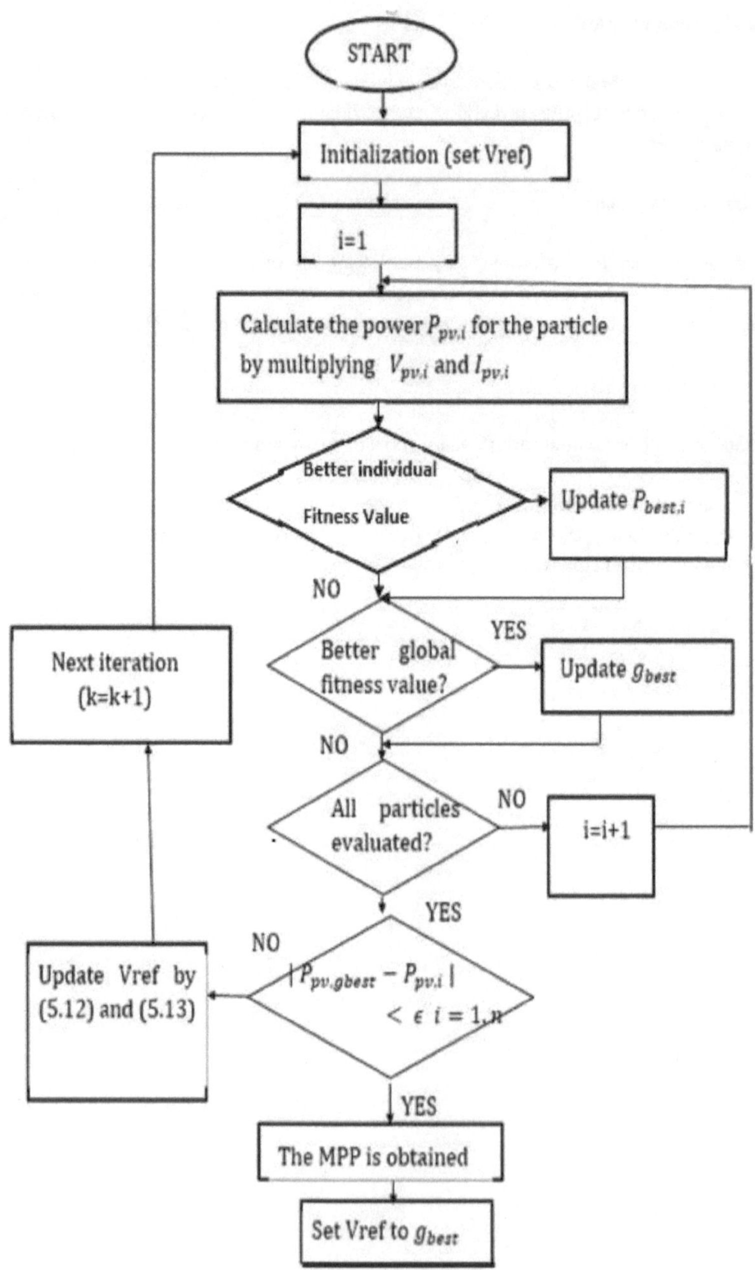

Figure 5.5:Particle swarm optimisation flow chart for MPPT[89]

5.3 Summary

The chapter explains the PSO and FA stochastic optimization techniques and how they are applied in PV MPPT. The FA has two tuning parameters that allows it to search for global optimization these are gamma (γ), it characterizes the variation of the attractiveness and it plays a very crucial role on how fast the FA converges. For most applications, it usually changes from 0.1 to 10. The other parameter alpha (α) allows the search process to escape the local search and search on the global scale.
The PSO has three parameters the inertia weight (w), cognitive component (c1) and the social component (c2). In the literature it was found that time varying of the PSO parameters produces more accurate results. In this book time varying parameters of the PSO were used.
 The convergence of the PSO and FA relies heavily on good selection of these parameters.
In PV MPPT the voltage control method is suggested. This method includes additional circuitry of a PI controller. Both the FA and PSO are optimising the best voltage value as the variable solution that will produce the best PV output power. The objective function to be optimised is the PV power.

Chapter 6: Simulation Results

This chapter describes the implementation and testing of the proposed MPPT controls of PV power systems in MATLAB/SIMULINK environment. Five MPPT algorithms are implemented. The computational intelligence algorithms are shown to be superior compared to conventional algorithms.

6.1   System design

Figure 6.1 shows the SIMULINK model. To perform the tracking of maximum power, the different optimisation algorithms are implemented. As seen from the Figure 6.1 the voltage control method was used were the algorithm searches the best voltage value (Vref) which is the best voltage that would produce the best output power and sets as the reference voltage value and then the PV output voltage tracks this reference voltage value. The PI controller parameters are tuned on a trial and error basis. The output of the control signal is then compared with the high frequency signal to produce the PWM that can be used to switch the mosfet.

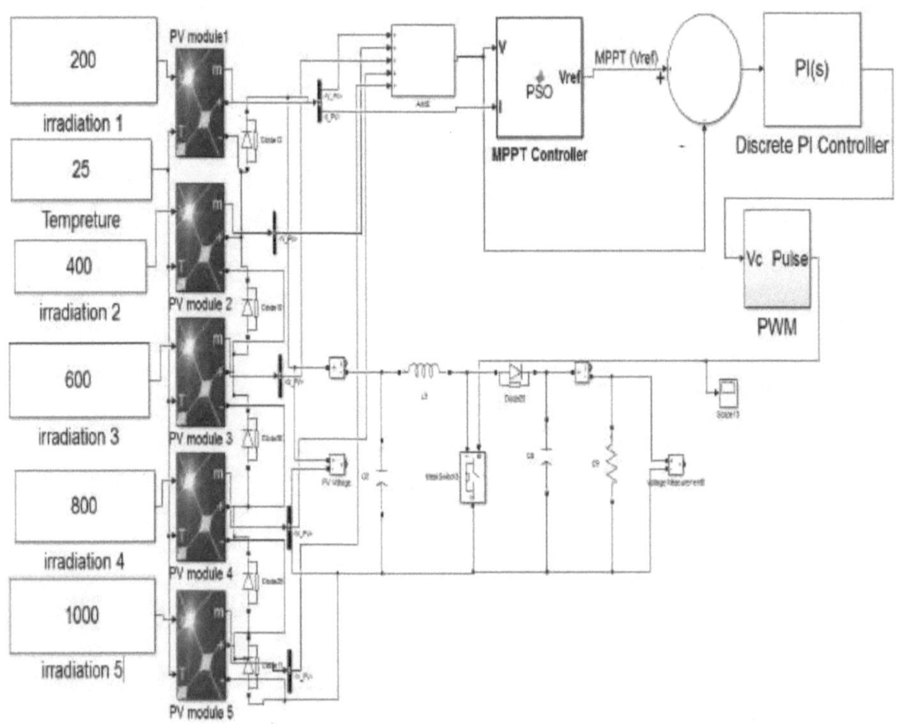

Figure 6.1 :Complete PV system

6.2 PV modelling and validation at STC

A Built-in Simulink PV model the 1soltechl sTH-215-P was used to verify its parameters, this built in model is from an actual PV module (see appendix A). Table 6.1 shows the parameters of the built-in module. These parameters were used to create a block diagram (see appendix B) of the module using the general PV model equations (2.1) - (2.7). Figure 6.2 shows the block in Simulink.

Table 6. 1:SIMULINK Built in PV module Parameters

Parameters	Values
Maximum power (Watts)	213.15
Open circuit Voltage V_{ov} (Volts)	36.3
Short circuit current I_{sc} (Amps)	7.84
Maximum power point Voltage V_{mp} (Volts)	29
Maximum power point current I_{mp} (Amps)	7.35
Temperature coefficient of V_{oc} (%/deg.C)	-0.36099
Cells per module	60
Temperature coefficient of I_{sc} (%/deg.C)	0.102
Photocurrent (Amps)	7.8649
Reverse saturation current (Amps)	2.9259^-10
Diode ideality factor	0.98
Series resistance R_s (ohms)	0.39
Shunt resistance R_{sh} (ohms)	313.40

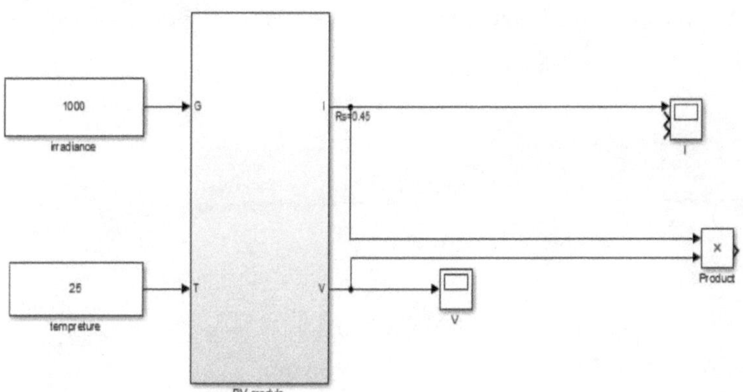

Figure 6.2:The built PV module

Figure 6.3 and Figure 6.4 show the results of the P-V and I-V curves respectively of the variation of Rs to obtain the maximum power on the data sheet. The initial Rs and Rsh were from the Matlab built-in PV module. Rsh was kept constant at 313.40 ohms whist changing Rs.

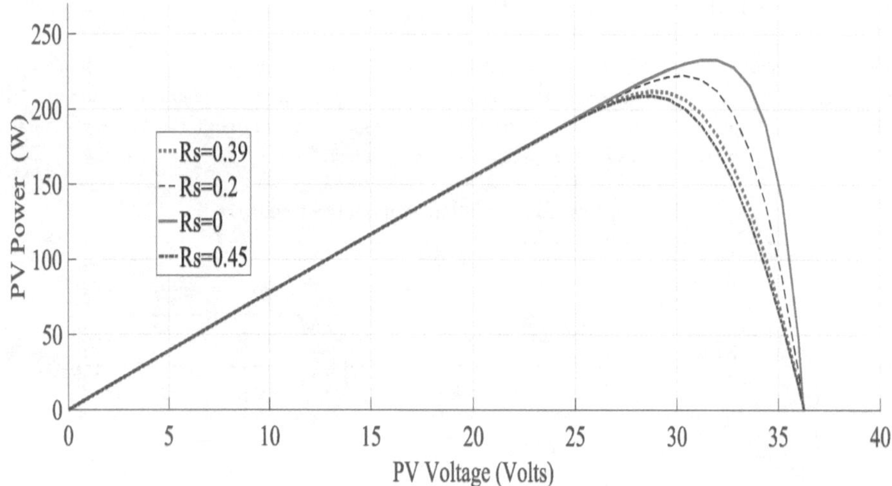

Figure 6.3: P-V curve of Variation of Rs with Rsh constant

From the curves it can be seen that the parameters of Rs and Rsh shown in Table 6.1, used in the built-in Matlab module do not produce the stated output power of 213.15 W instead it produces 212.1 W. Rs at 0.45 ohms produced 209.1W,Rs at 0.2 ohms produced 222.6W and Rs at 0 ohms produced 233.1W. The variation of Rs is consistent with literature i.e as Rs reduces the power increases or vice versa as can be seen in Figure 2.9.

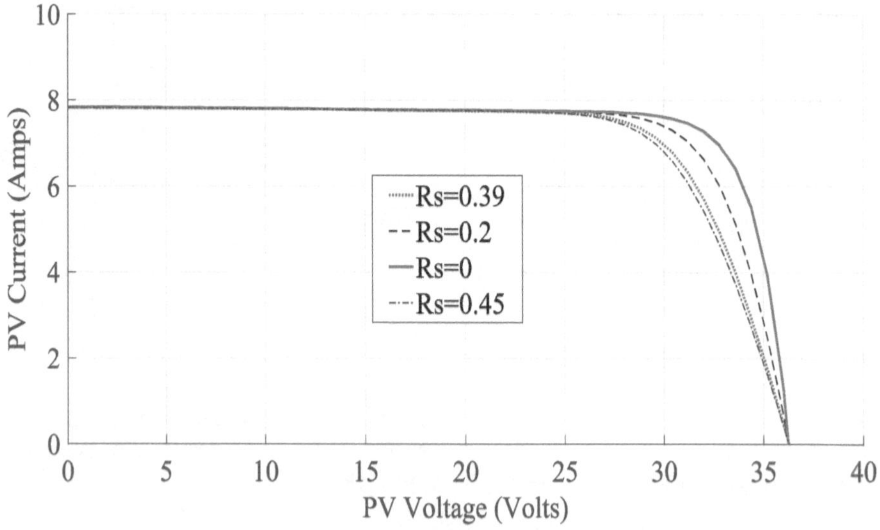

Figure 6.4: I-V curve of variation of Rs with Rsh constant

Now Rs is kept constant at 0.39 ohms and Rsh value is varying to see its effect on power and current. Figure 6.5 and Figure 6.6 show the P-V and I-V curves respectively of the results.

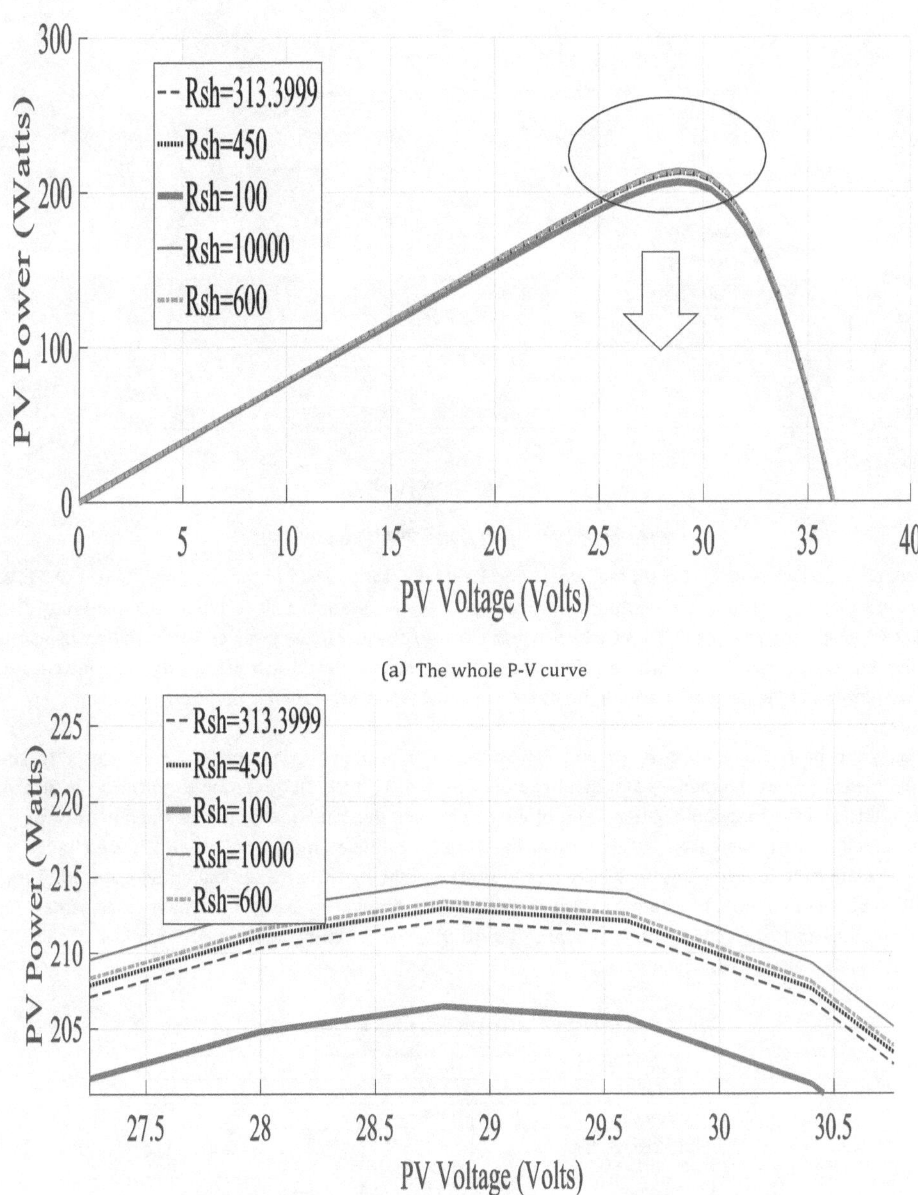

(a) The whole P-V curve

(b) Zoomed at MPP

Figure 6.5:P-V curve varying Rsh with Rs constant

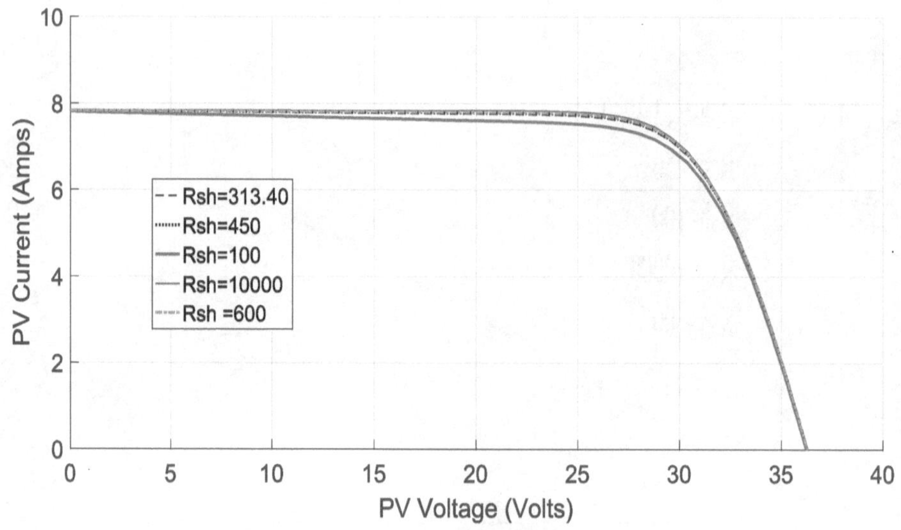

Figure 6.6:I-V curve varying Rsh with Rs constant

It can be observed in Figure 6.5 (b) that when Rsh=10000 ohms it is closer to the datasheet power of 215W (appendix A) as it produces an output power of 214.7W. The Matlab built in model parameters of Rsh =313.40 yields a power of 212.1 W which is not close to the data sheet value. The variation of Rsh is consistent with the literature where it is suggested that an increase in Rsh will result in an increase in power until Rsh reaches a value where the power will not change anymore if it keeps on increasing.

Finding the best combination of Rs and Rsh to give the best approximation of power closer to the experiment value of the module datasheet can be a challenging task. Optimisation algorithms can also be used to find the best combination. It is not the objective of this book to find the best parameters for the module. The Matlab built in PV model seen in Figure 6.1 with parameters of Table 6.1 is used for all the simulations to follow in this book to reduce loop errors that exist when a self-built model is used. The Matlab built in PV model of Figure A. 2(Appendix A) is connected in series with a bypass diode across each module to form a PV string as shown in Figure 6.1. The PV string consists of 5 modules. Table 6.2 shows the calculations of the string.

Table 6. 2:PV String of 5 modules in series

Module Parameter (STC)	Value
Maximum power (W)	213.15×5 =1065.75
Maximum voltage (V)	29×5=145
Current at max power (A)	7.35×1=7.35
Open circuit voltage (V)	36.3×5=181.5
Short circuit current (A)	7.84×1=7.84

Figure 6.7 and Figure 6.8 show the P-V and I-V curves of the string whilst varying temperature and keeping irradiance constant at $1000W/m^2$.

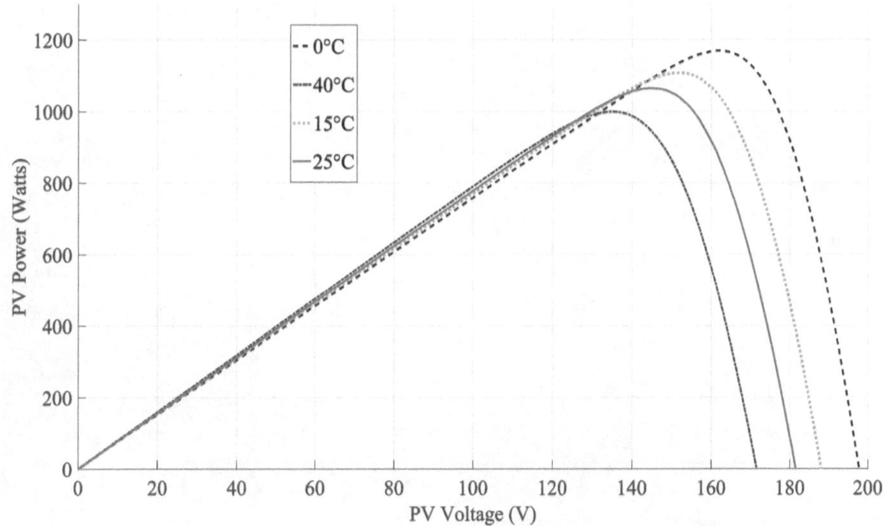

Figure 6.7:P-V curve showing variation in tempreture with irradience constant

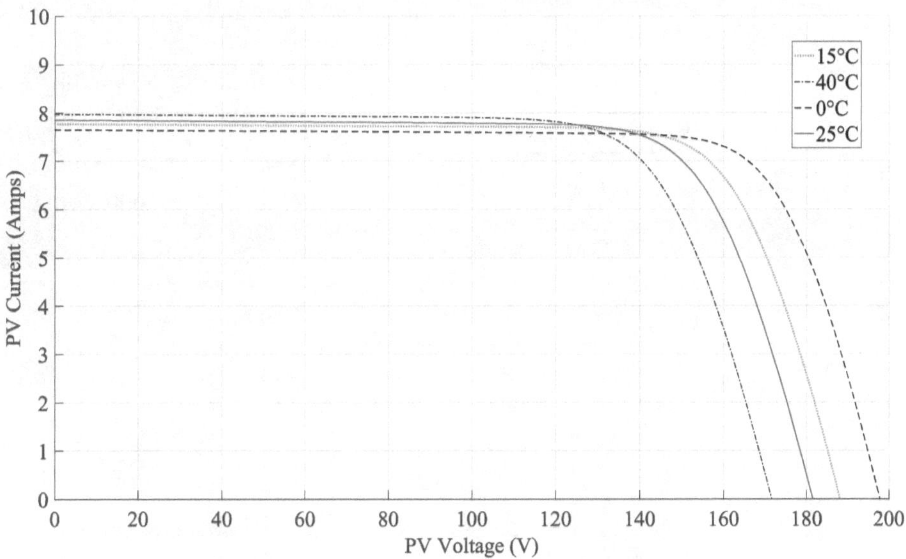

Figure 6.8:I-V curve of variation of tempreture with irradience constant

The simulation results are consistent with literature.

Figure 6.9 and Figure 6.10 show the P-V and I-V curves with temperature kept constant at 25 °C and irradiance varying.

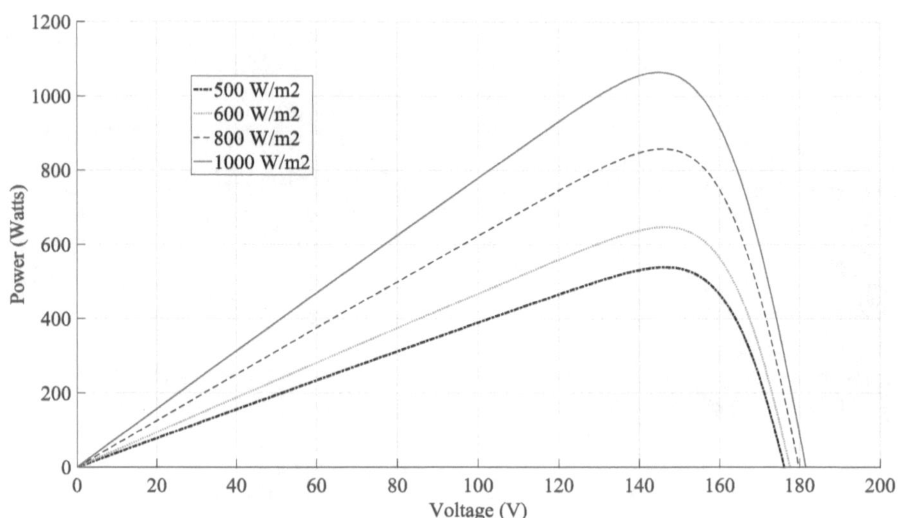

Figure 6.9:P-V curve of variation in irradience with tempreture constant

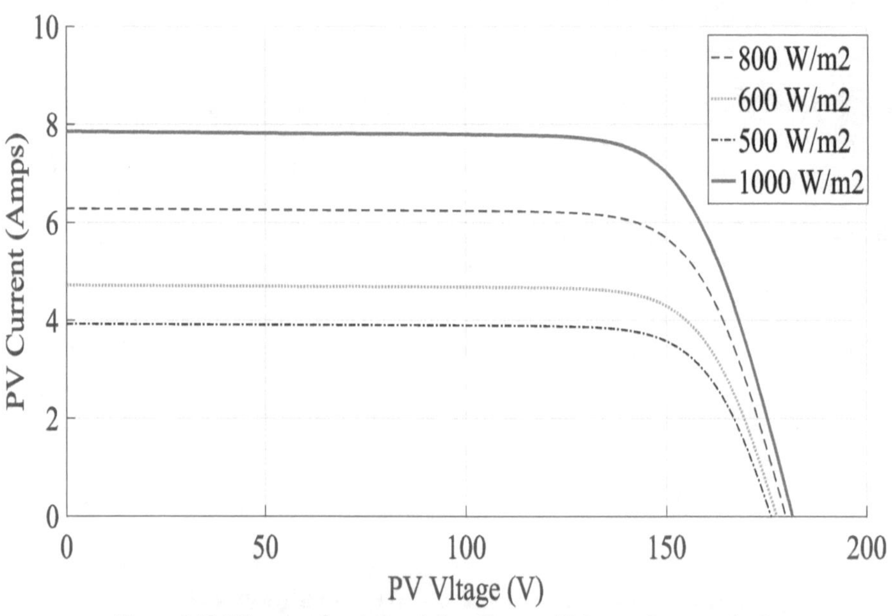

Figure 6.10:I-V curve of variation in irradience with tempreture constant

The results are consistent with literature where the PV power increases with increase in irradiance and also as irradiance increases the short circuit current increases.

6.3 Boost Converter Design and Control

The boost converter is modelled in Simulink. The circuit diagram is shown in Figure 6.11. The boost converter was tested with a DC source and a pulse generator to see if it performs as expected. Table 6.3 shows the parameters of the boost converter and Table 6.4 shows the boost converter parameters that were calculated using equations (2.16), (2.17), (2.19) and (2.20).

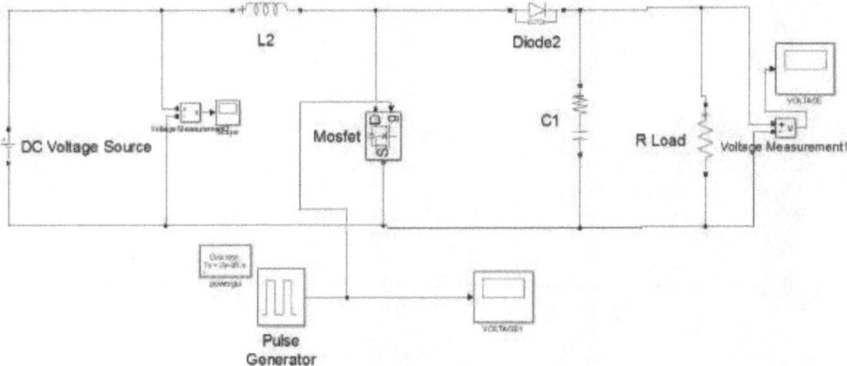

Figure 6.11:Boost converter

Table 6.3:Boost converter specification

Specifications	Value	unit
Input Voltage (PV voltage at STC)	145	V
Input current (PV current at STC)	7.35	A
Output voltage	300	V
Input Current ripples (3%) ΔI_L	0.2205	A
Voltage output ripple (2%) ΔV_{out}	6	V
Input power (output power of PV at STC)	1065.75	W
Switching frequency	20	kHz

Table 6.4:Boost conerter Parameters

Calculated	Value	Unit
Duty cycle, D	51.600	percent
Load resistance , R_o	84.440	Ω
Output current, I_o	3.552	A
Inductance, L	0.017	H
Capacitance ,C	$1.5273e^{-5}$	F

Figure 6.12 shows the simulation results of the converter and Figure 6.13 shows the zoomed results showing how the voltage is changing.

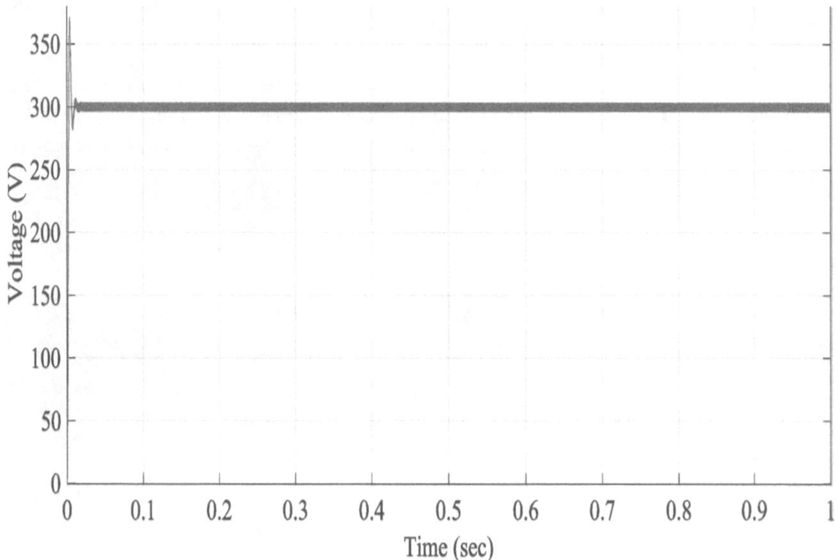

Figure 6.12: Boost conveter output voltage

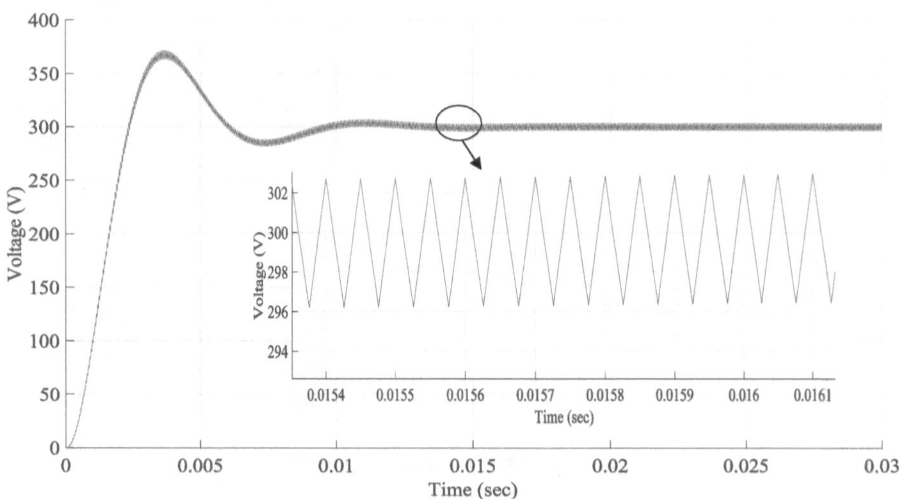

Figure 6.13: Zoomed boost conveter output voltage

The results show that the converter is performing as expected according to the design specifications.

The state space small signal model of the boost converter discussed in chapter 4 is implemented in the Matlab mfile to produce the transfer function. The parameters of the boost of R, L and C are the same as seen in Table 6.4. The mfile code to get the small signal model is given in appendix C. After getting the transfer function classical methods such as the Ziegler Nicolas is used to find the controller parameters that would give the best closed loop response in terms of a short settling time, small overshoots, zero steady state error and an appropriate rise time. The transfer function obtained is shown in equation (6.1).

$$\frac{V_0}{d} = \frac{-4.8091 \times 10^{\wedge}5\ (s - 1168)}{(s^2 + 775.3s + 9.057 \times 10^{\wedge}5)} \tag{6.1}$$

It can be seen that it is a second order system, this is expected because of the two variables that exist. There is a zero on the right of the s plane which agrees with literature. Figure 6.14 shows the root locus of the system.

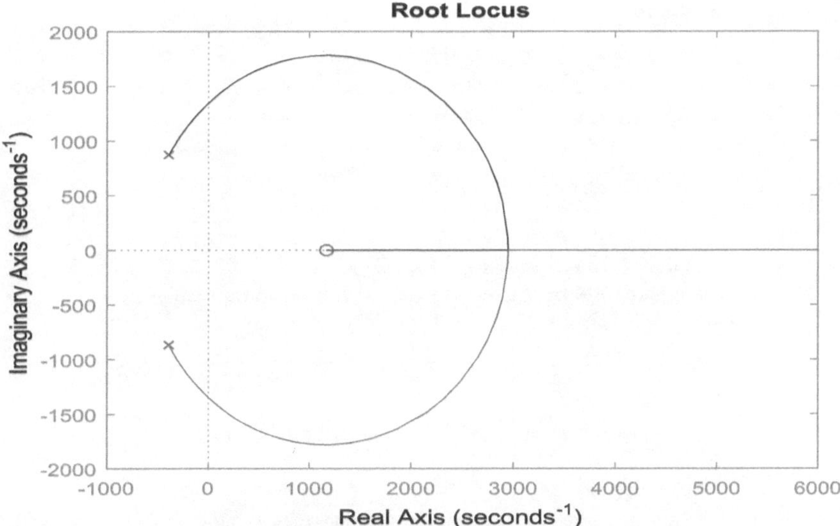

Figure 6.14: Root Locus of the Boost Converter

It can be seen that there are two complex poles on the left of the s plane.

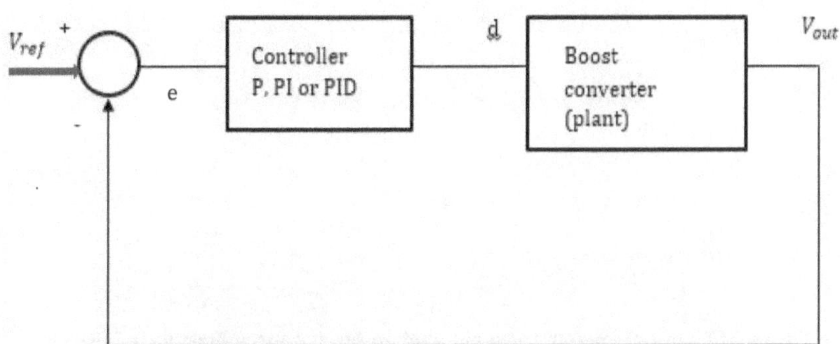

Figure 6.15:Closed loop control of the boost conveter

The purpose of the controller is to make the error (e), that is, the difference Vref and Vout to be zero. It should be noted that the d signal (duty cycle) in Figure 6.15 is limited between 0 and 1. A PI controller was used in the closed loop system. The SISO tool in Matlab was used to find the parameters of P and I. The Ziegler Nicholas step response was used until a stable response was found. Figure 6.16 shows the Matlab SISO tool with the boost converter transfer function. Equation 6.2 shows the PI controller G(s).

$$G(s) = K_p + \frac{K_i}{s} = \frac{K_i(1 + s\frac{K_p}{K_i})}{s} \qquad (6.2)$$

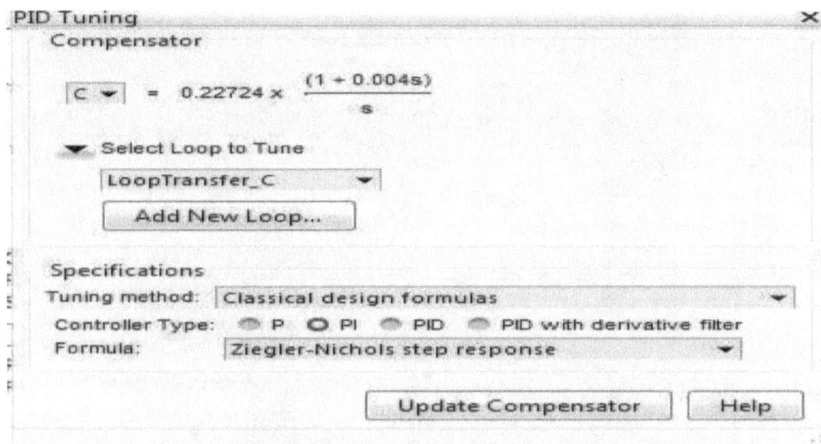

Figure 6.16:PI parameters found using the sisotool

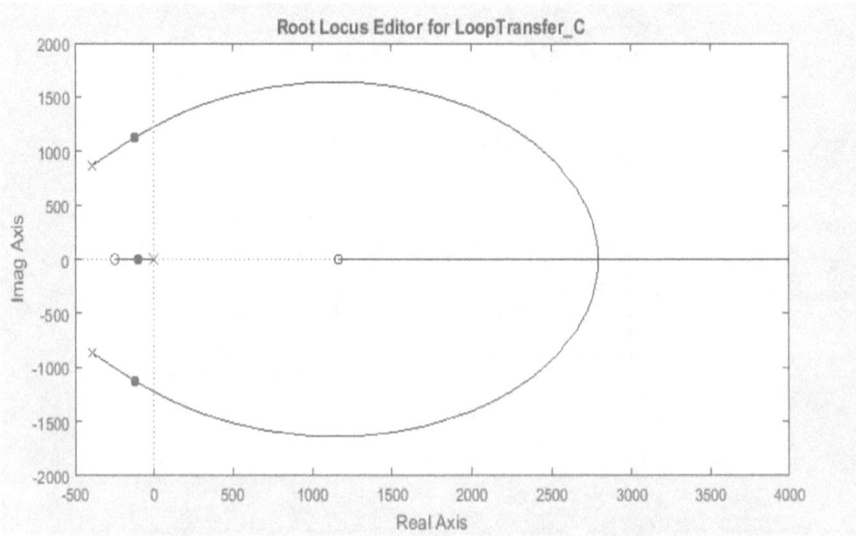

Figure 6.17 :Closed loop root Locus editor of the boost transfer function in SISO tool.

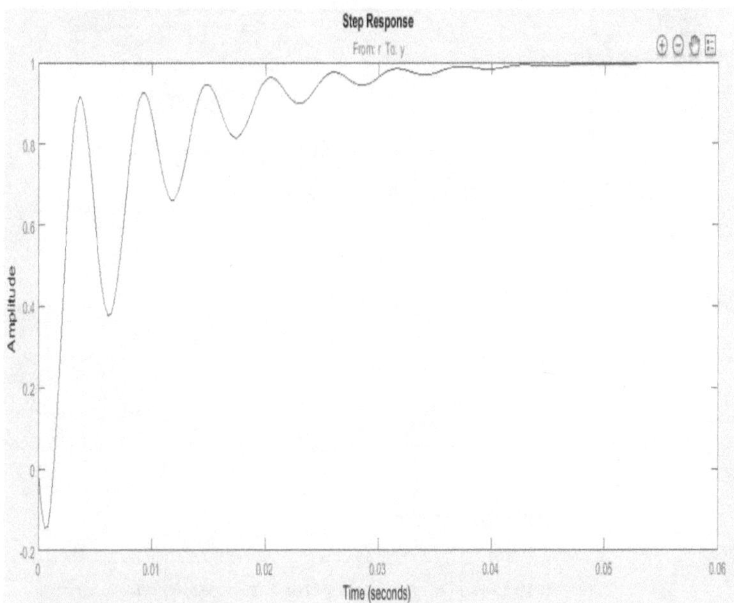

Figure 6.18:Corosponding step response of the transfer function in sisotool

The SISO tool was used to find a stable response. The closed loop poles shown in Figure 6.17 were moved along the loci until a stable step response shown in Figure 6.18 was found. The corresponding P and I gains can be found by matching equation 6.2 and the values found in Figure 6.16. The PI parameters calculated are shown in equations 6.3 and 6.4.

$$K_i = 0.227 \tag{6.3}$$

$$K_p = K_i \times 0.004 = 0.00090896 \quad = 9.0896e^{-4} \tag{6.4}$$

These parameters were also tested in a Matlab mfile with the corresponding transfer function and the closed loop response is shown in Figure C. 1 (Appendix C). These parameters make the model stable but it should be noted that this is a linear model of the real system of the boost converter so the hope is that these parameters can be used on the actual system (large signal) and perform relatively the same. Figure 6.19 shows the boost converter in closed loop with the controller. The PI parameters found were inputted in the PI block to see how the actual system would perform. A reference of 200 V was used as a set point to see if it would be tracked. Then different step references of 100V, 250V and 150V were used. Other Ki and Kp values were used to show the robustness of the controller. Figure 6.20 and Figure 6.21 show the results. The PWM block is used to compare the control signal with the 20 kHz high frequency signal to produce the pulse that switches the mosfet.

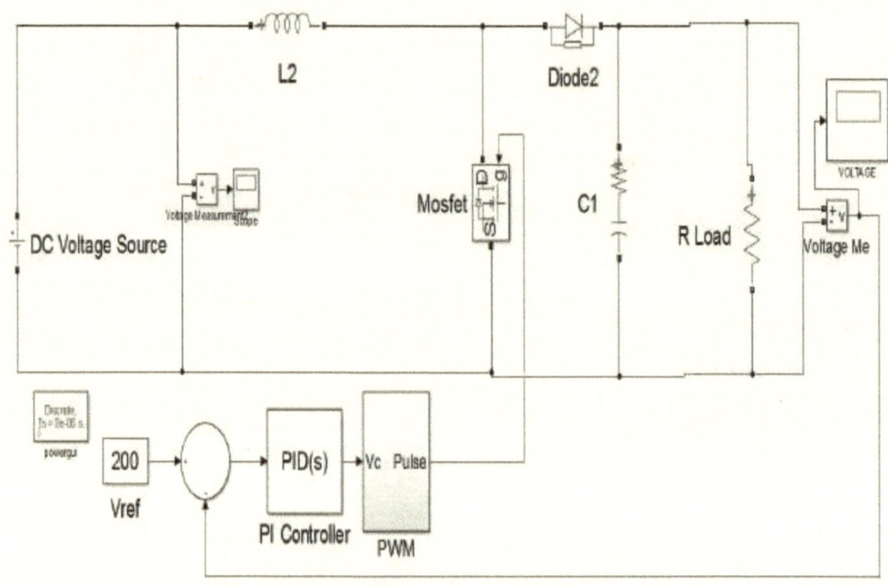

Figure 6.19:Closed loop system of the Boost converter

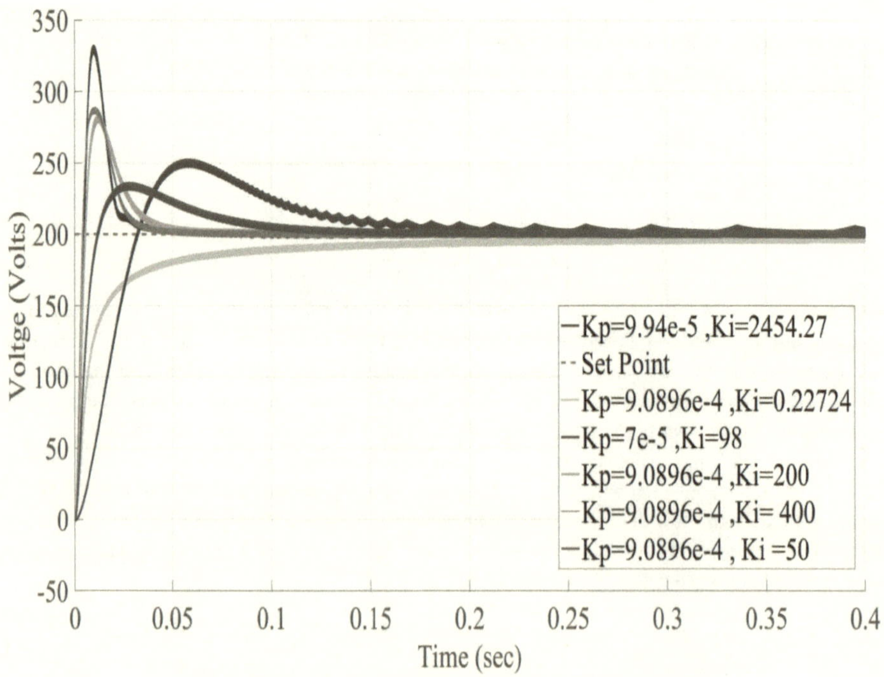

Figure 6.20: PI controller performance with a refrence of 200V

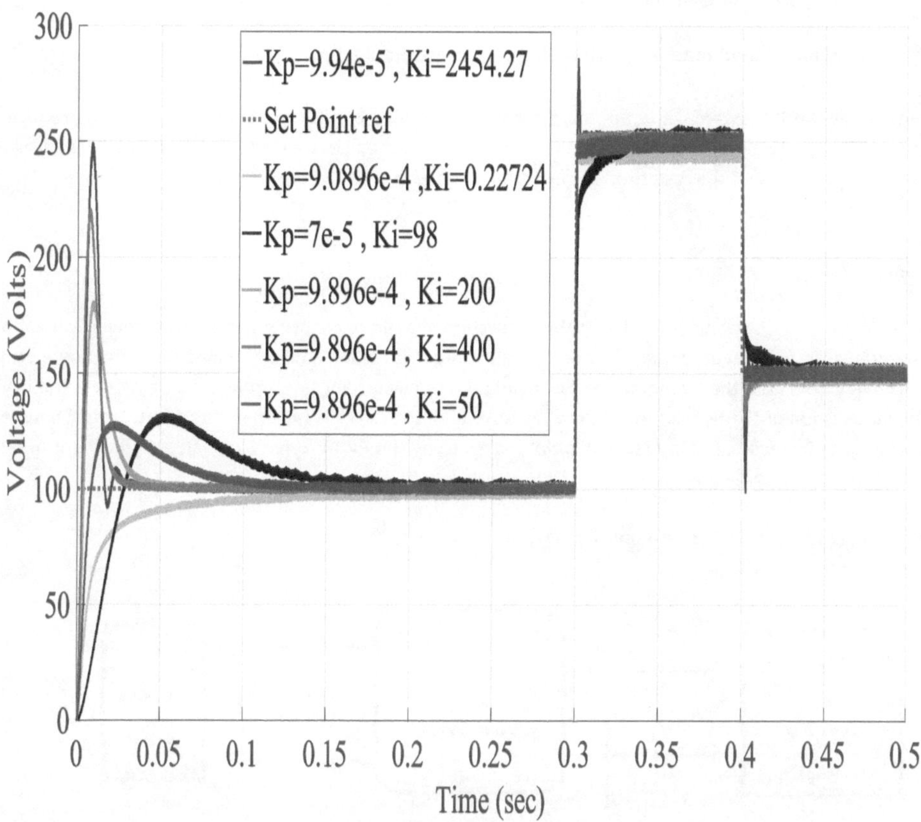

Figure 6.21 :PI controller performance with step refrences

The above results show that different PI parameters will produce different performances. There are infinite solutions that can be obtained so the idea is to find parameters that work according to performance specifications. From Figure 6.20, the parameters found using the SISO tool can be seen as not been able to fast track the set point, these parameters gave a guideline of where other solutions can be found. The Kp gain is made to remain constant at $9.0896e^{-4}$ and Ki gain was varied. It can be observed that a Ki gain of 200 gives a better performance than the one calculated by the SISO tool which is 0.227.

Figure 6.21 shows that the set point ref is able to be tracked for the different step changes. It should be noted that optimum PI parameters can also be found by defining a mathematical multi objective function then using optimisation algorithms to find the best parameters according to the function. However, this book is not focusing on this area.

6.4   MPPT using conventional algorithms and metaheuristic algorithms

In this section five optimisation algorithms are compared under different atmospheric conditions for maximum power point tracking in solar PV systems these are the perturb and observe (PnO), Incremental conductance (IC), Fuzzy logic (FL), Particle swarm optimisation (PSO) and the Firefly algorithm (FA). The complete system can be seen in Figure 6.1.

6.4.1 Principle of load matching according to the system design

Considering the PV system setup of 5 modules in series (Table 6.2) for maximum power to be produce a load of 19.727 ohms has to be used if a direct connection is used i.e. shown in equation (6.5). Assuming at STC. Appendix D shows the results of a direct connection with a 19.727 ohms load.

$$\frac{V_{mpp}}{I_{mpp}} = \frac{145}{7.35} = 19.727 \ ohms \tag{6.5}$$

So if a load of 84.440 ohms is used with this PV system in a direct connection maximum power will not be extracted as in our system design. Figure 2.16 illustrates that when a direct connection is made the same voltage is across both the PV module and load, and the same current runs through the PV module and load. The intersection of the two curves i.e., the I-V curve and load curve is the operating point. From Ohms law as resistance increases, the intersection point moves along the PV I-V curve from left to right therefore PV power will vary accordingly.

Figure 6.22 shows MPPT using a boost converter

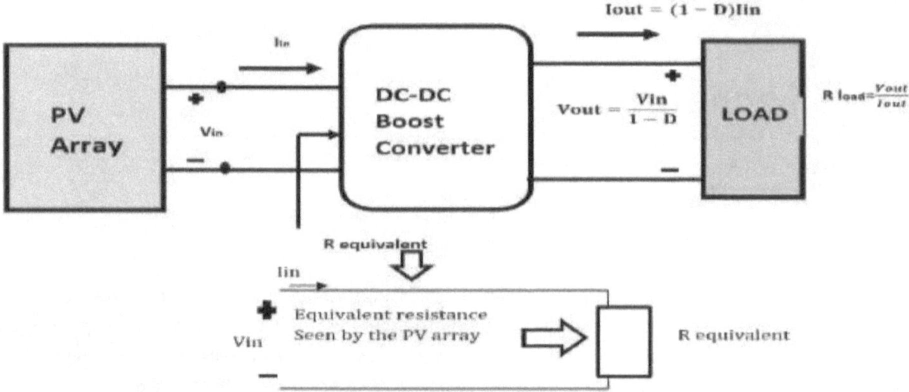

Figure 6.22: Connection of the Boost converter with the solar array

Using equation (2.16) and (2.17) the input resistance seen by PV module (*R equivalent*) is related to the output load impedance (*R load*) as shown in equation (6.6):

$$R_{equivalent} = \frac{V_{in}}{I_{in}} = \frac{(1-D)V_{out}}{\frac{I_{out}}{1-D}} = (1-D)^2 \frac{V_{out}}{I_{out}}$$

$$= (1-D)^2 R_{load} \tag{6.6}$$

By changing (D), *R equivalent* can be equated with the best load resistance at which the maximum power will be produced from the solar PV systems. Equation (6.6) is solved for duty cycle

$$D = 1 - \sqrt{\frac{R_{equivalent}}{R_{load}}} \qquad\qquad (6.7)$$

where $R_{equivalent}$ _is_ the same as equation (6.5). So in this setup the duty cycle becomes 0.516 as seen in Table 6.4.

6.4.2 Algorithm efficiency at static atmospheric conditions

The optimisation algorithms were tested under static irradiance with temperature kept constant at 25°C and the DC load is also kept constant at the rated value of 84.44 ohms. This was done to investigate the performances of algorithms in terms of steady state error and tracking efficiency. It should be noted that the P-V curve landscape is unimodal under static irradiance meaning that there is one local maximum power point (LMPP) which is also the global maximum power (GMP). It is expected that the deterministic algorithms i.e, PnO and IC should have no problem in tracking the MPP in this landscape.

A population of reference voltages (Vref) is created for the stochastic algorithms. Table 6.5 shows the parameters used for the algorithms for optimum GMP searching. A step size of 2V was selected for the PnO and IC. For the fuzzy logic the step size varies according to how far $\Delta E(k)$ is from the MPP and is calculated by the rule base Table. Appendix E shows the fuzzy logic setup of inputs and outputs. The optimal parameters selection for the PSO and FA were obtained based on a concept from the literature [30], [83], [84], [85]. The optimal parameter selection of (α) and (γ) for the Firefly algorithm and (c1, c2 and w) for the PSO were selected based on running a series of tests to find out which values converge close to the optimum value. Initial selection of these parameters was obtained by testing benchmark functions of Sphere, Ackley, Rosenbrock, Rastrigin and Griewank. The Max iteration of 40 was selected on the basis that convergence had occurred. The population of 7 was selected to reduce the number of iterations when solving the problem. The search space was between 80V and 180V, this search space was selected based on knowledge of the likely location of the optimum solution. The random Vector of Vref voltages selected was [137V, 130V, 110V, 140V,125V, 135V, 150V] this was based on the vicinity of where the MPP is likely to occur. For PnO, IC and Fuzzy logic the starting point ($V_{PV}(k-1)$) of the perturbation was selected to be 80% of the open circuit voltage V_{oc} , this was found to be where the MPP voltage is likely to occur. The direct connection of the design load (84.44 ohms) to the PV system was also investigated to find out how much power is lost.

Table 6.5:Optimisation algorithm parameters

Parameters	PSO	FA	PnO	Fuzzy Logic	IC
Population	7	7			
Max iteration	40	40			
w_{max}	1				
w_{min}	0.1				
$c1_{max}$	2				
$c1_{min}$	1				
$c2_{max}$	2				
$c2_{min}$	1				
α		0.2			
γ		1			
Step size			2V	variable	2V
Boundaries	80V-180V	80V-180V			
Dimension	1	1	1	1	1
Variable	voltage	voltage	voltage	voltage	voltage

A series of simulations were performed to find out how much power can be extracted with the algorithms under different constant irradiance with temperature kept constant at 25°C. Figure 6.23 shows the PV power results of the simulation at an irradiance of 1000 W/m2 i.e., STC for the PnO and IC. The MPP indicated on every Figure represents the theoretical maximum power point. Figure 6.24 shows the corresponding voltage and Figure 6.25 shows the current.
Figure 6.26 shows the extracted power by the PSO, FA and Fuzzy logic. Figure 6.27 and Figure 6.28 show the corresponding PV voltage and current respectively.

Further simulations were performed for an irradiance of 800, 600 and 500 W/m2 to investigate the performance of the algorithms. The theoretical maximum power (MPP) can be seen on the P-V curves of Figure 6.9. To calculate the efficiency equation (4.1) is used. Table 6.6 shows the complete results of the performance of the algorithms. The (800 W/m2, 600W/m2 and 500W/m2) simulation results of the PV power extracted can be seen in Appendix F. Figure 6.29 shows the convergence speed of the algorithms at STC. Speed of convergence was not compared as this depends on where the initialisation of Vref begins. The energy lost due to a direct connection is also calculated and can be seen in Table 6.6.

Figure 6.30 shows the search process of the FA to find the best voltage (Vref) that gives the best PV power. The PV voltage takes time to converge with the Vref because of the PI parameters. Fine tuning of this PI controller can be a challenging task.

For the Figures indicated, a potion was zoomed in to make it clear. The circle in the Figures indicates were the zoomed in section was done.

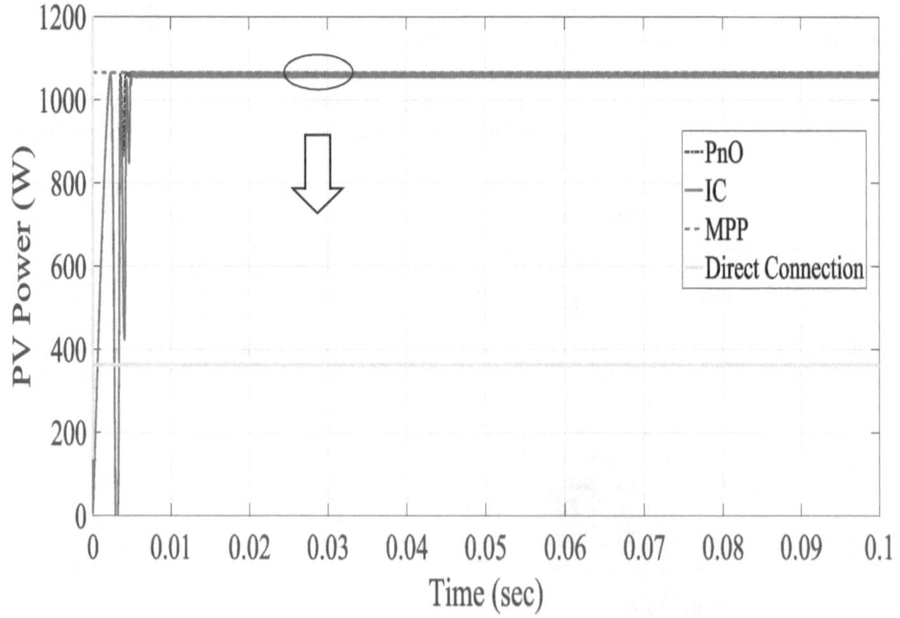

(a) MPP tracking for 0.1s duration

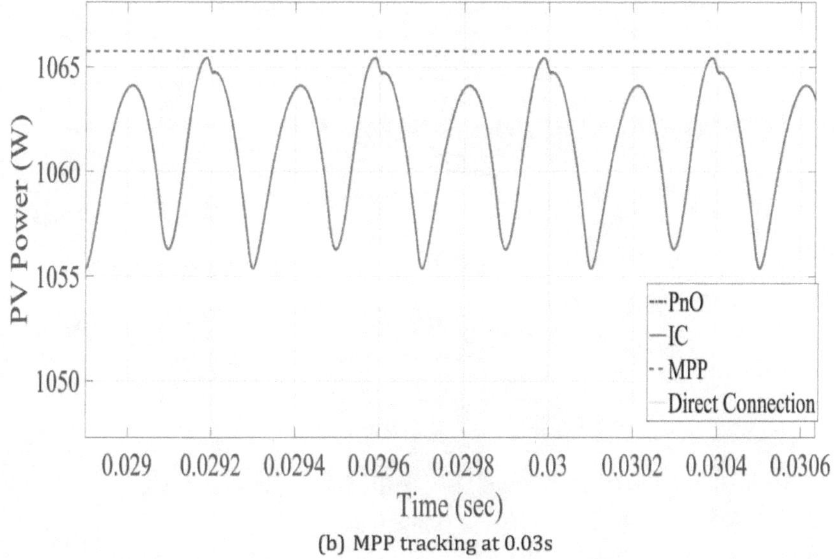

(b) MPP tracking at 0.03s

Figure 6.23: Extracted PV power under an irradience of 1000 W/m2 by the PnO and IC

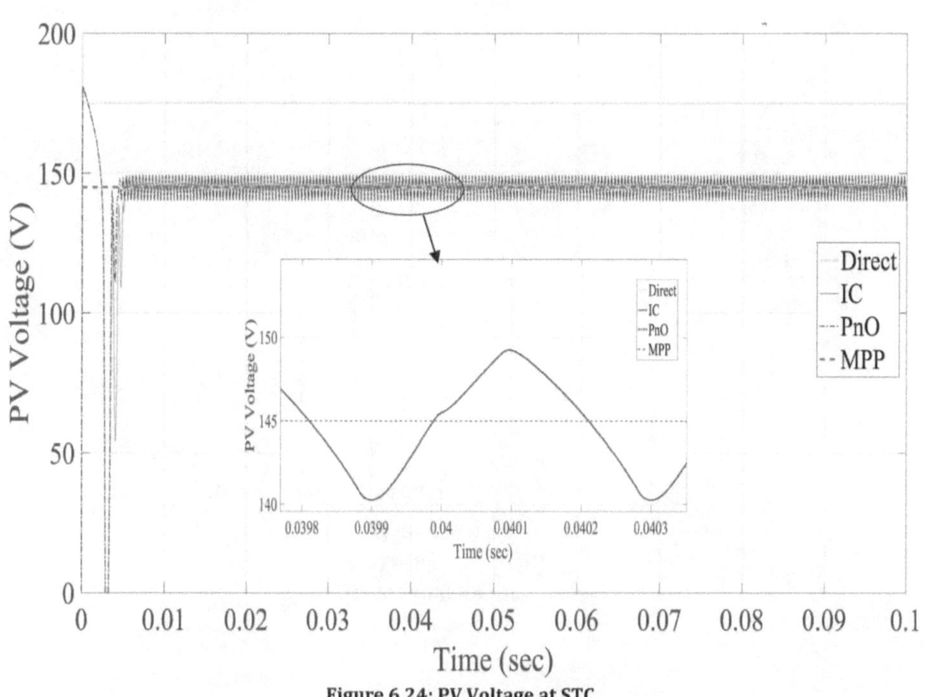

Figure 6.24: PV Voltage at STC

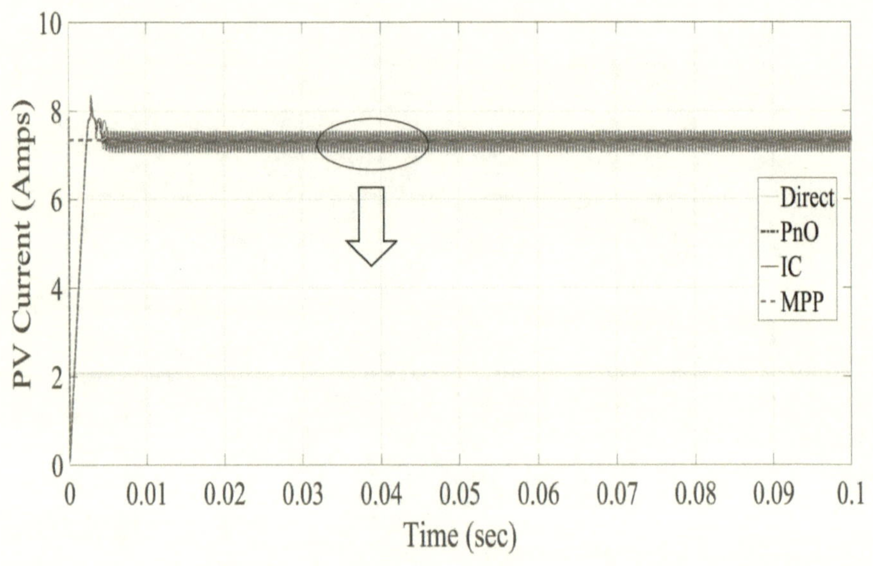

(a) PV current for 0.1s duration

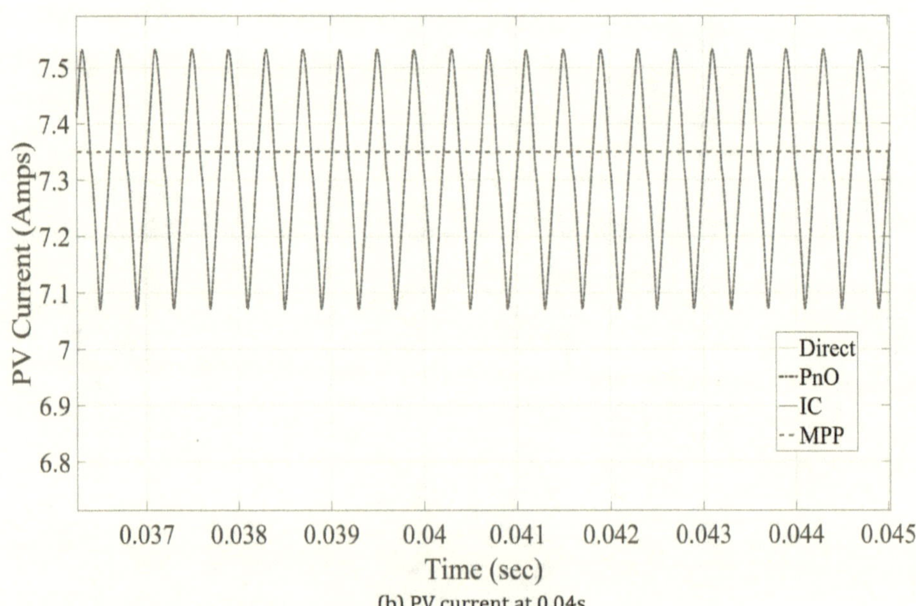

(b) PV current at 0.04s

Figure 6.25: PV Current at STC

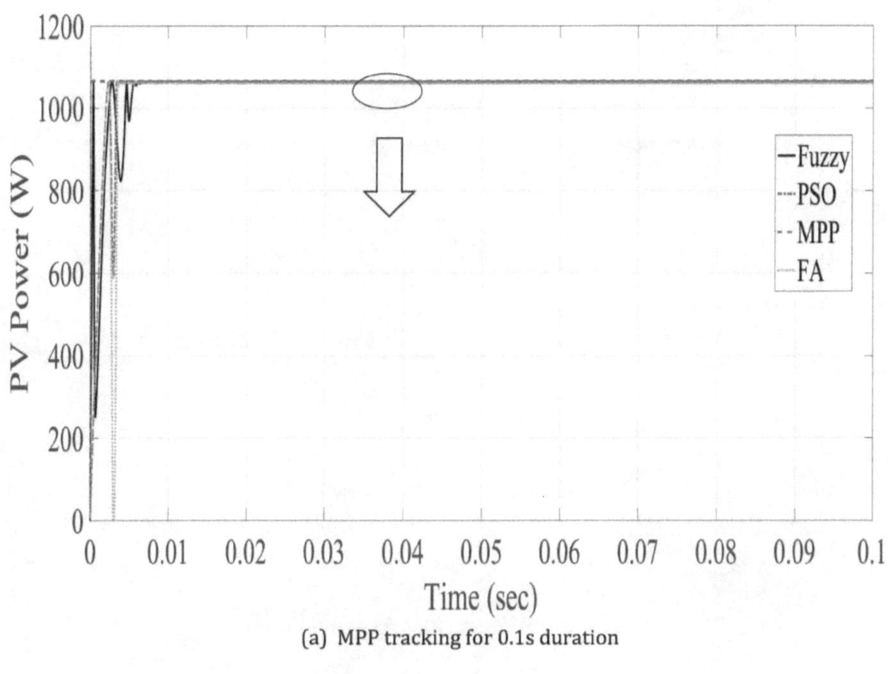

(a) MPP tracking for 0.1s duration

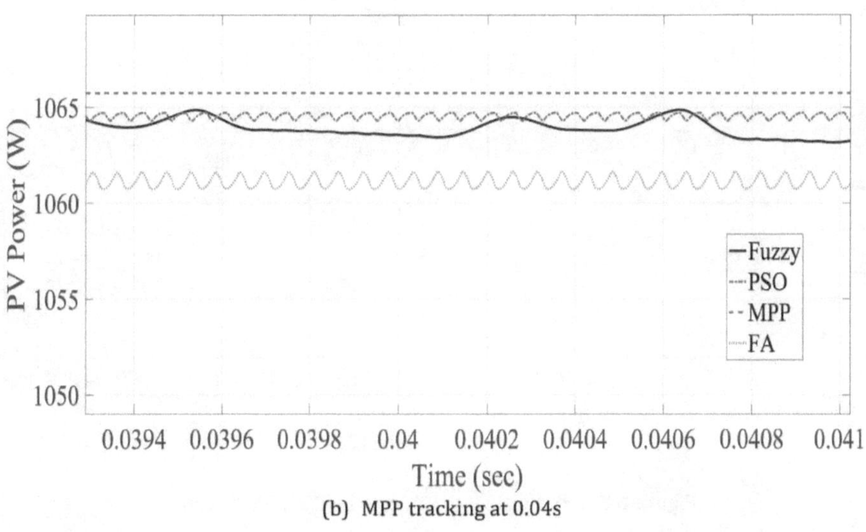

(b) MPP tracking at 0.04s

Figure 6.26: Extracted PV power under an irradiance of 1000 W/m2 by the PSO, FA and Fuzzy

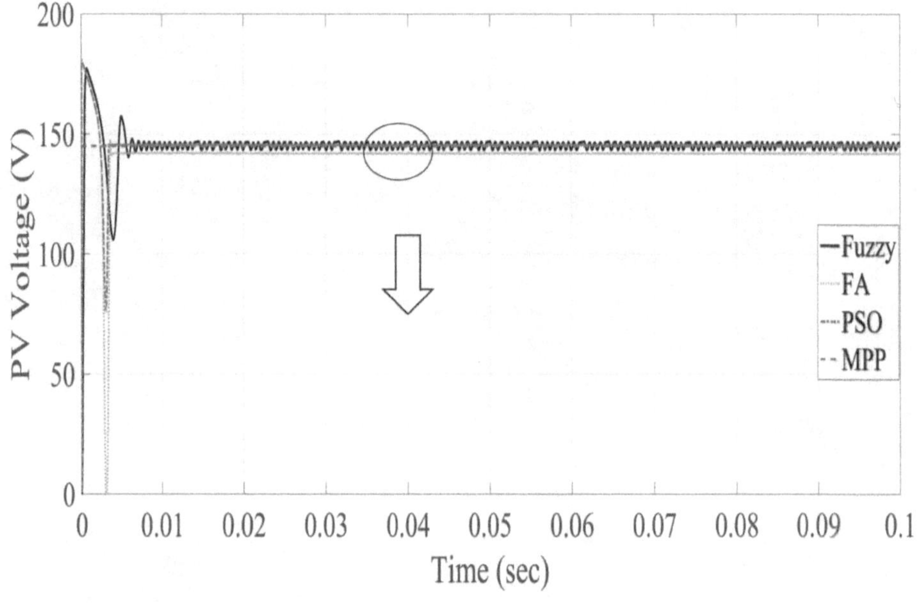

(a) PV Voltage for 0.1s duration

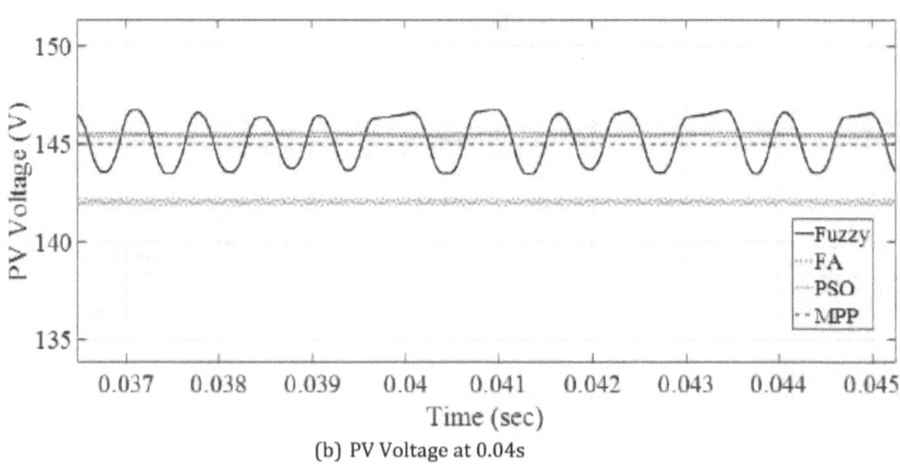

(b) PV Voltage at 0.04s

Figure 6.27: PV Voltage at STC for the PSO, FA and Fuzzy

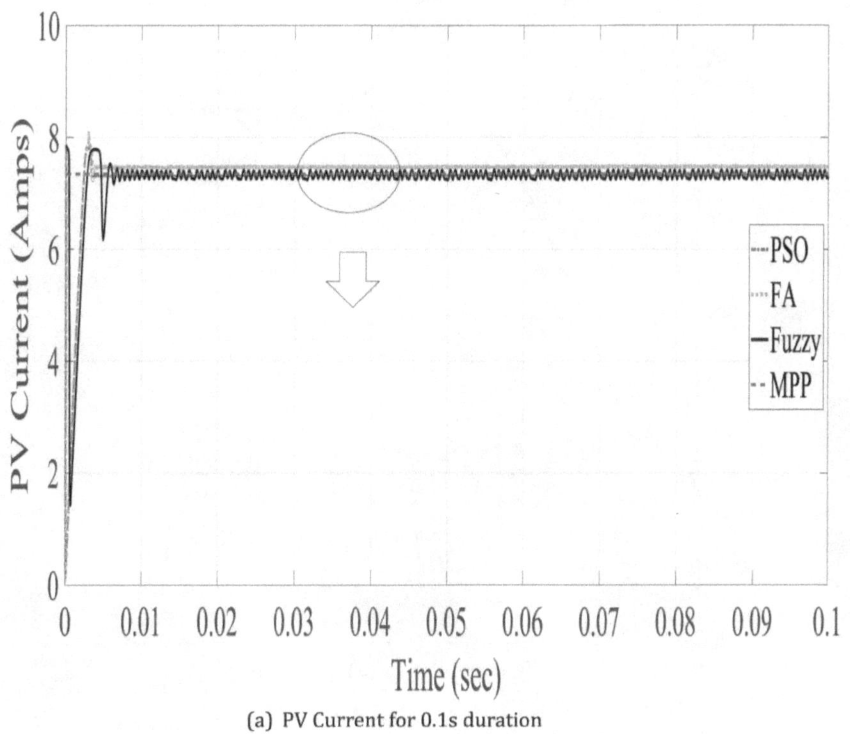

(a) PV Current for 0.1s duration

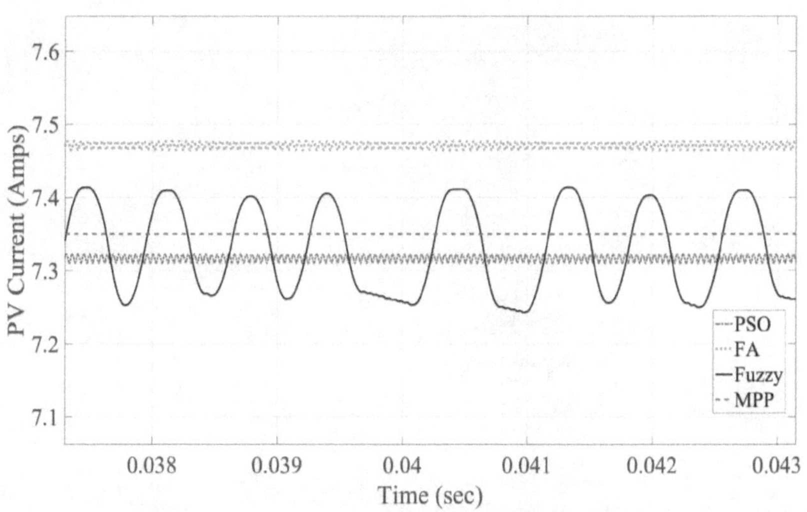

(a) PV Current at 0.03s

Figure 6.28 PV current at STC

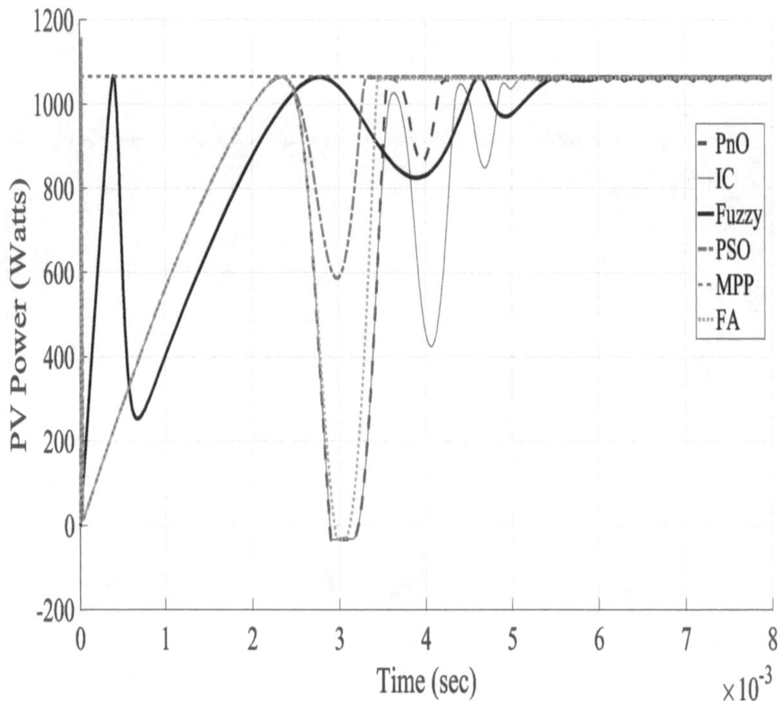

Figure 6.29: Convergency speed of algorthms at STC

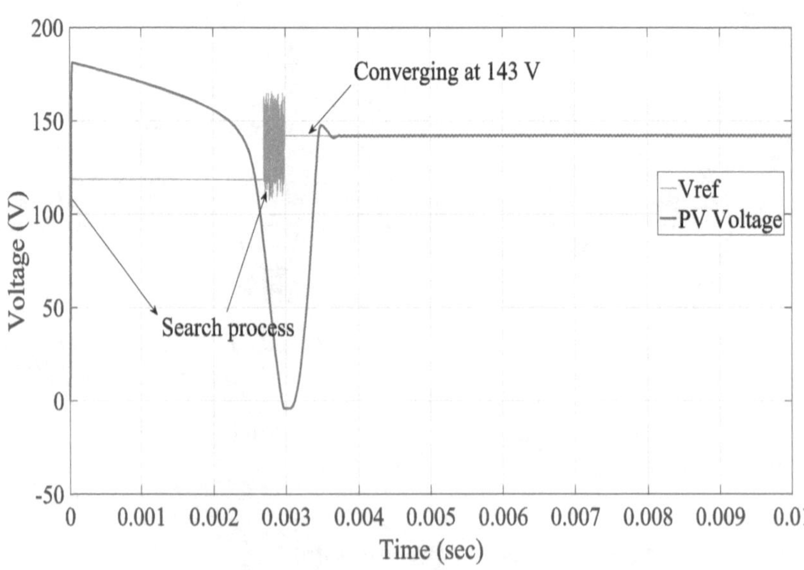

Figure 6.30 :Search process of the FA in finding best voltage (Vref)

Table 6.6:Performance of the algorthms at different static irradience

Irradiance(W/m^2)		1000	800	600	500
Theoretical P_{max} from PV curve(watts)		1065.75	858.9	647.5	539.9
PnO MPPT	P_{mppt} (W)	1060	850	642	538
	Efficiency (%)	99.5	98.9	99.2	99.6
IC MPPT	P_{mppt} (W)	1060	852	643	538
	Efficiency (%)	99.5	99.2	99.3	99.6
Fuzzy MPPT	P_{mppt} (W)	1064	858	646	538.5
	Efficiency (%)	99.8	99.9	99.8	99.7
PSO MPPT	P_{mppt} (W)	1064.5	858.9	647	539.5
	Efficiency (%)	99.8	100	99.9	99.9
FA MPPT	P_{mppt} (W)	1062	858.9	646.5	539.5
	Efficiency (%)	99.6	100	99.8	99.9
Direct connection Power (W)		363	353	340	330
Power loss due to Direct connection (%)		65.9	58.9	47.5	38.9

Table 6.6 shows that all algorithms were able to track at static atmospheric conditions at high efficiencies from high insolation to low insolation levels. As expected, the PnO and IC ring once steady state has been reached (Figure 6.23 (b)) this is because of the constant step size of 2 V. The ringing of the Fuzzy Logic is minimum because it creates a variable step size depending on how far it is from the MPP, the step size is according to the rule base table. The CI based stochastic algorithms have the least amount of ringing at steady state this is expected because of their nature of searching. For the PSO, once the optimum Vref has been found the velocity of the particles reduces close to zero hence there is minimum ringing. For the FA once the flies converge to the brightest firefly there is less ringing. It can also be seen that a substantial amount of power is lost if a direct connection is made due to the impedance mismatch between the PV module and the load. Equation (6.8) was used to calculate the power loss due to direct connection.

$$\text{Power loss due to direct connection (\%)} = \frac{Theoretical\ Pmax - Direct\ connection\ power}{Theoretical\ Pmax} \times 100 \qquad (6.8)$$

6.4.3 Performance of the five algorithms under rapid varying irradiance

A rapid varying solar irradiance profile was used to test the performances of the algorithms. The temperature was still kept constant at 25°C. Figure 6.31 shows the irradiance profile used and Figure 6.32 shows the results of the PV power tracked by the PnO and IC. The corresponding PV voltage and PV current are shown in Figure G. 1 and Figure G. 2 of appendix G. It should be noted that all five modules in the string received the same irradiance at the same time so the P-V curve landscape is still unimodal. It should also be noted that because the direct connection is not viable and adequate, it was not included in the rest of the simulations.

Three points were zoomed in from Figure 6.32 (a) these are (b), (c) and (d). It can be observed that the PnO lost track of the MPP at time 0.36s in Figure 6.32(b) and time 1.2s to 1.6s in Figure 6.32 (c). This is due to the fact that the algorithm fails to distinguish the power increases, whether it is from perturbation or a change of irradiance. As the irradiation increases, the operating point moves away from the MPP.

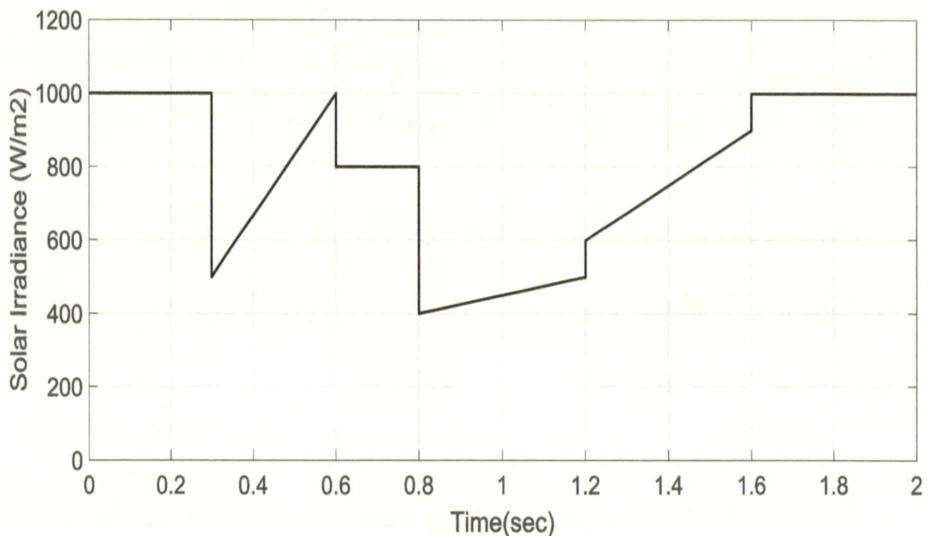

Figure 6.31 :Varying Irradience profile

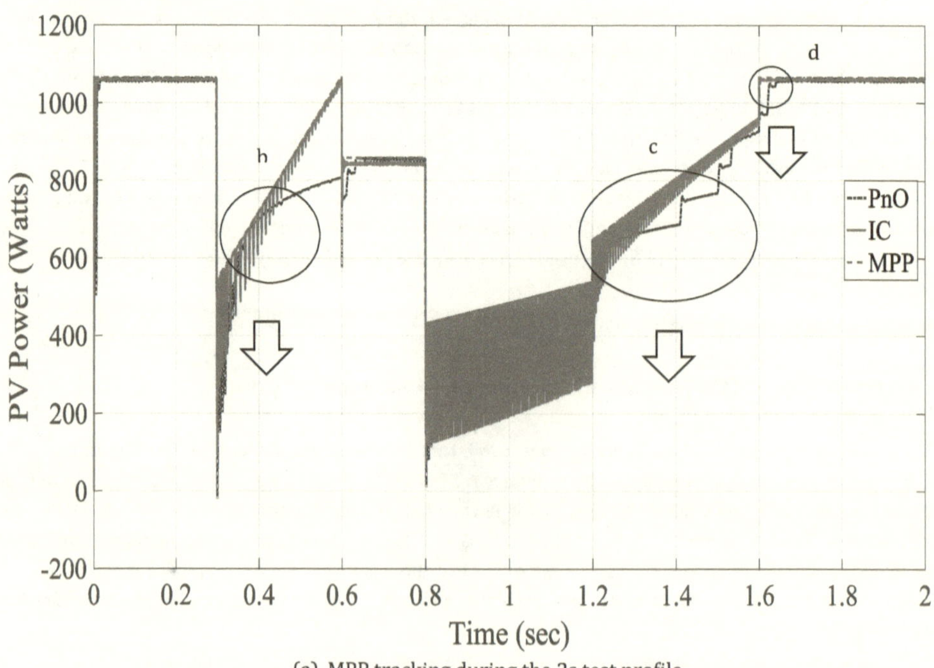

(a) MPP tracking during the 2s test profile

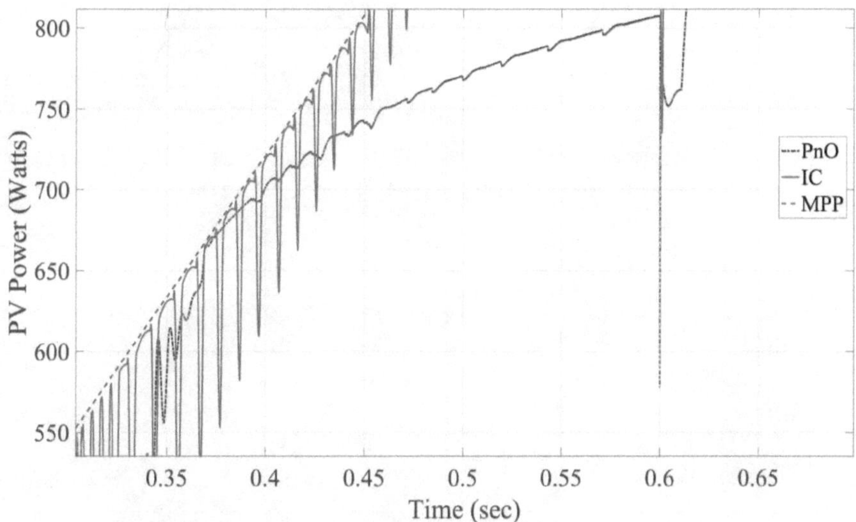

(b) MPP tracking at fast change in solar irradiance at 0.35s

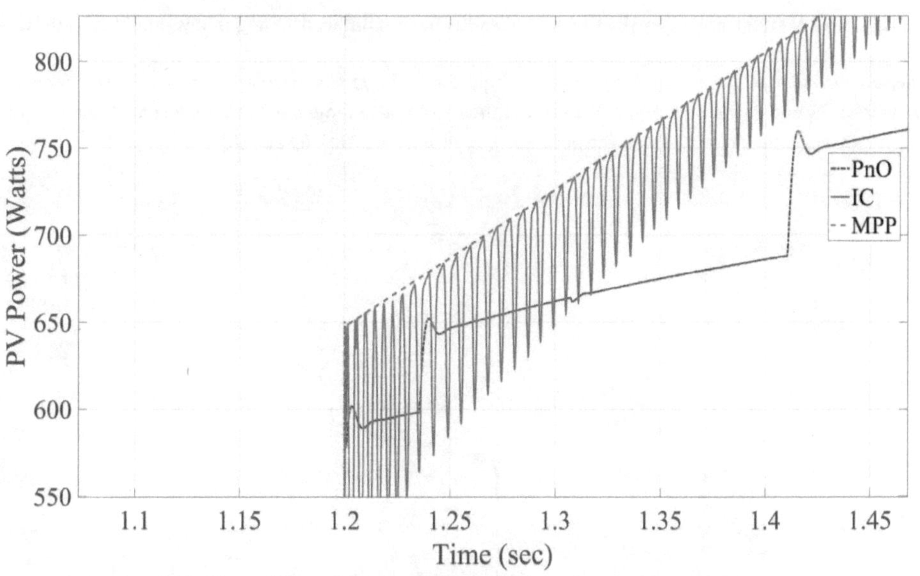

(c) MPP tracking at fast change increase in solar irradiance at 1.2s

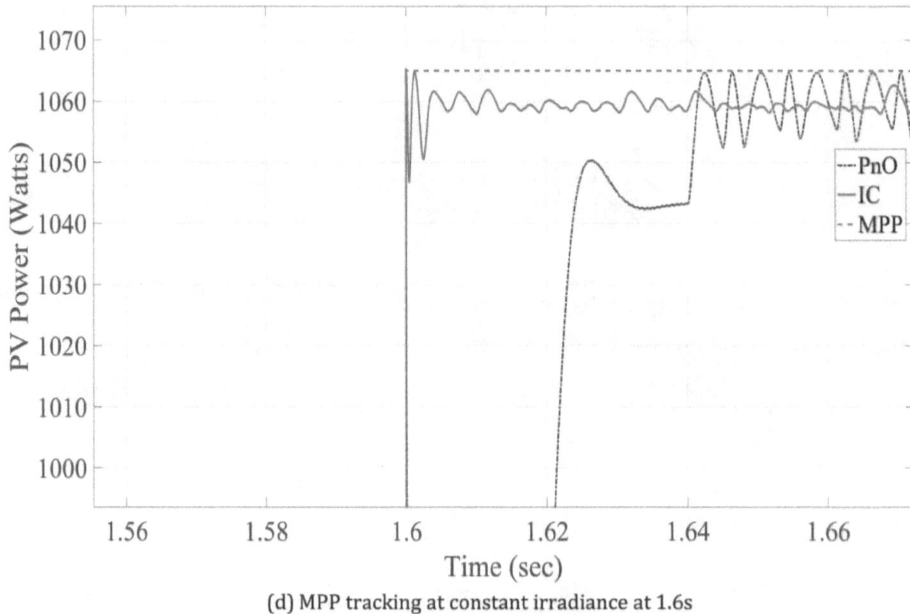

(d) MPP tracking at constant irradiance at 1.6s

Figure 6.32: MPP tracking by the PnO and IC controllers at different changes in weather conditions

The same irradiance profile of Figure 6.31 was used to test for varying irradiance for the CI algorithms that is the PSO, FA and Fuzzy logic. Figure 6.33 shows the results. The same points as the one for the PnO and IC results were zoomed in to examine the difference, see Figure 6.33 (b), (c) and (d).

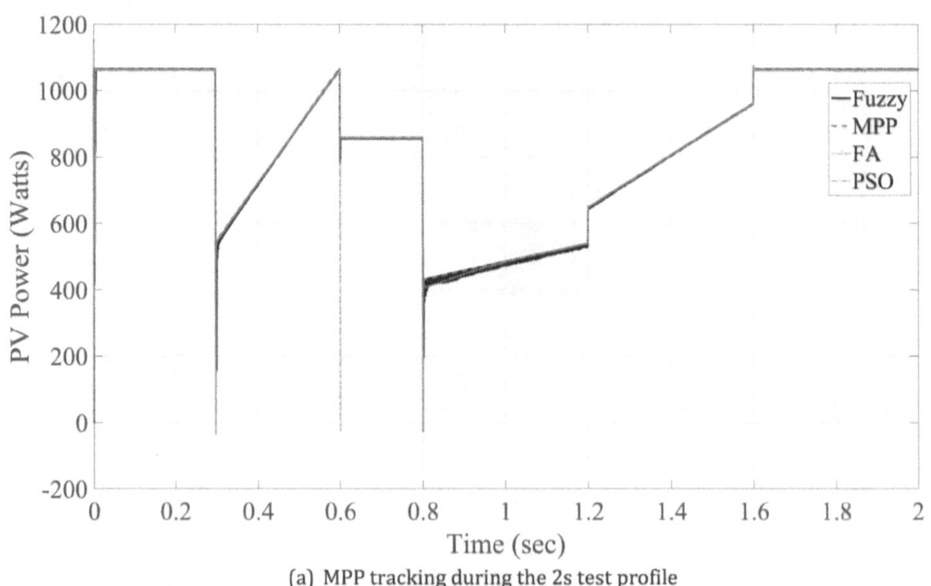

(a) MPP tracking during the 2s test profile

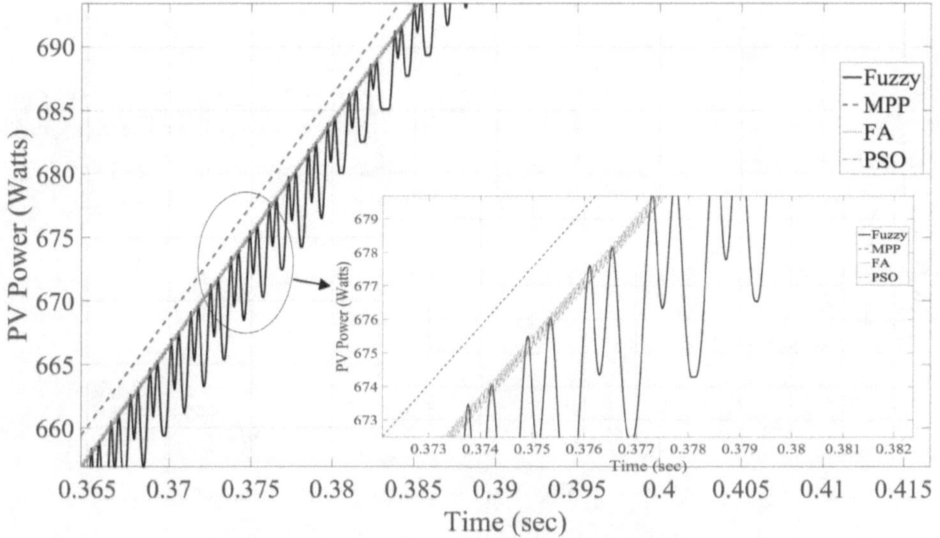

(b) MPP tracking at fast change in solar irradiance at 0.35s

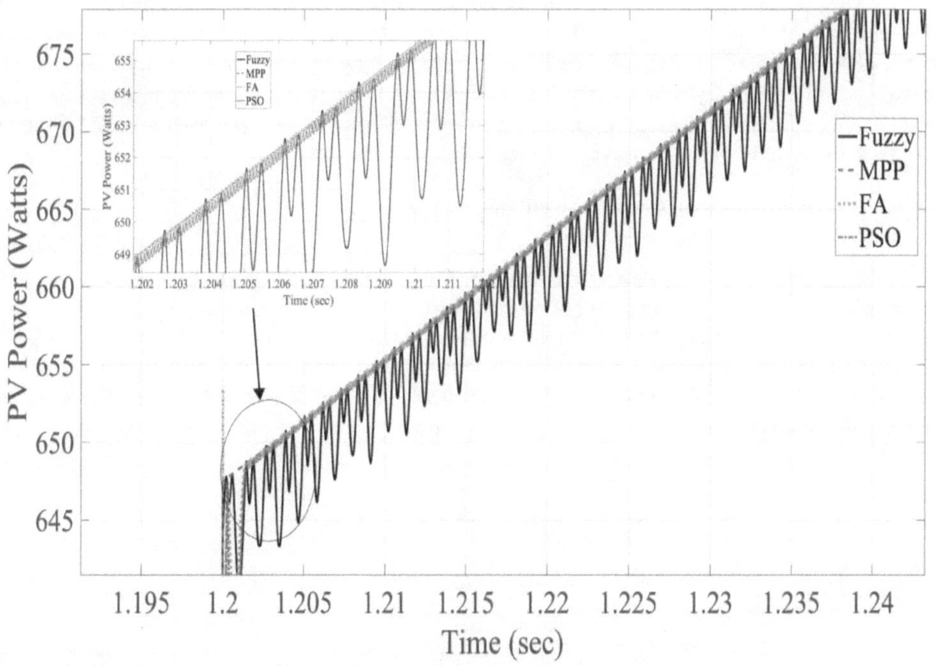

(c) MPP tracking at fast change increase in solar irradiance at 1.2s

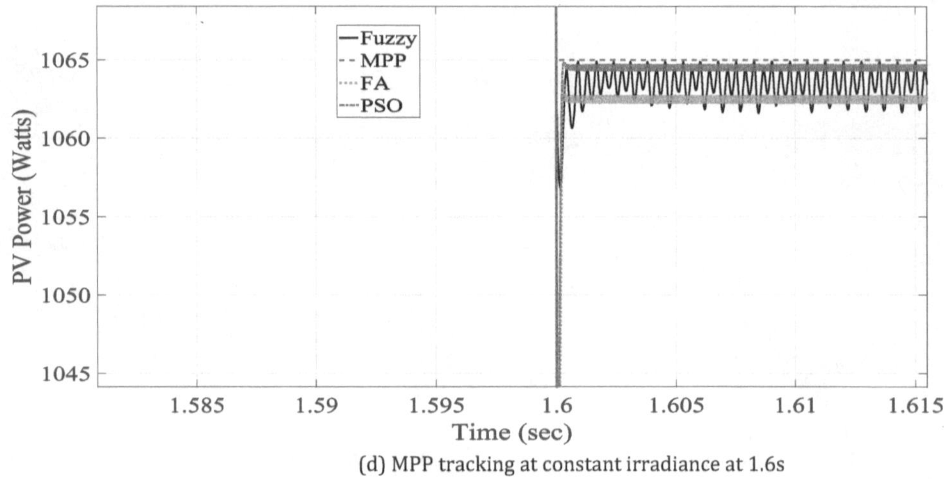

(d) MPP tracking at constant irradiance at 1.6s

Figure 6.33: MPP tracking by the PSO, FA and Fuzzy controllers at different changes in weather conditions

The power loss caused by this misinterpretation of the PnO algorithm is not significant this is because the MPPT is fast enough. The IC tracked the MPP without getting lost but it can be observed in Figure 6.32 (a) that the ringing at steady state becomes less with increases in irradiance. The PSO, FA and Fuzzy were able to track the MPP without having any complications as seen in Figure 6.33 (a). The fuzzy logic, PSO and FA tracked accurately with minimum ringing and followed the theoretical MPP as close as possible. This suggests that these controller give the best performance. PSO and FA have the least amount of ringing compared to the other algorithms this can be seen in Figure 6.33 (b), (c) and (d).

6.4.4 The effects of the step size for the PnO algorithm

`The PnO algorithm requires a step size to find the MPP. A trade-off is required between accuracy and speed. In this section, different step sizes of perturbation are used to show that a big step size will tack the MPP faster but will ring more at steady state whilst a small step size is slow but will rings less at the MPP. For this study, four step sizes were chosen that is, 0.3V, 1V, 3V and 5V. Figure 6.34 shows the PV power results. The test was done at a constant irradiance of 1000 W/m2 and 25 °C i.e. STC. Figure 6.35 shows the speed of the different step sizes. As the results indicate an appropriate trade- off has to be chosen.

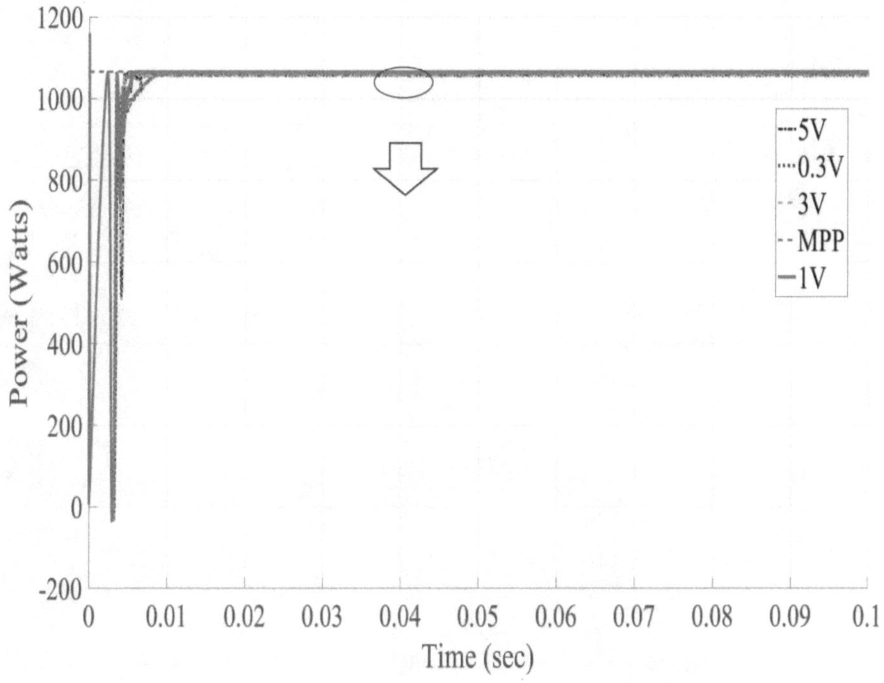

(a) MPPT of different step sizes of the PnO for 0.1s

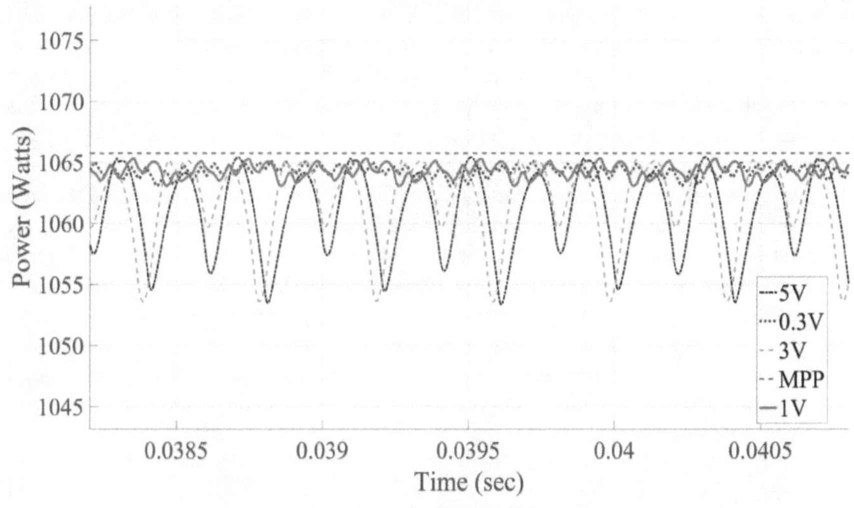

(b) MPPT at 0.04s

Figure 6.34: Effects of different step size for the PnO under STC

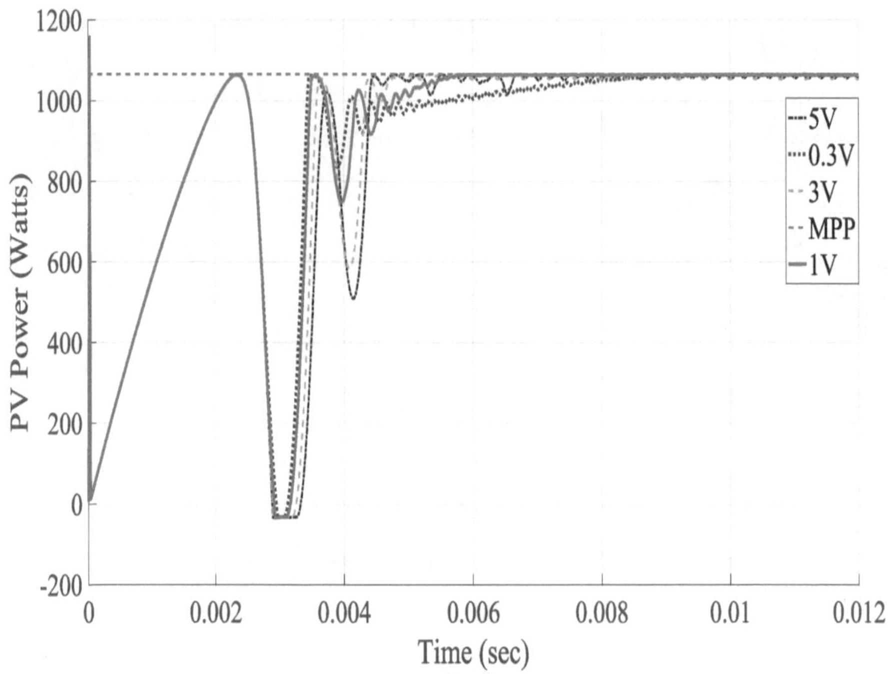

Figure 6.35: Convergency speed of the different step size

6.4.5 Partial shading conditions

For the static irradiance and the varying irradiance tests the P-V curve landscape has been unimodal and the superiority of the stochastic optimisation was not seen other than less oscillations at steady state as all algorithms were able to find the MPP. Under partial shaded conditions (PSC), the P-V curves now becomes multimodal the ability of the stochastic algorithms to search locally and globally in a multimodal function will be investigated. The PnO will be used to show why a gradient based algorithm is not reliable in tracking the GMPP under a multimodal function. As mentioned in the optimisation chapter for deterministic algorithms, they are very sensitive to the starting point. The PnO will be made to start at different points close to a local peak or global peak this is done to see if it will converge to the peak it's close to. The IC is also a gradient based algorithm similar to PnO, so it is not included in this comparison. The Fuzzy logic is usually implemented with another algorithm like artificial neural networks, PSO or other global searching algorithms to form a hybrid for it to find the GMPP in a multimodal environment so it was not considered.

Five PSC patterns are produced by exposing the PV at different irradiance on each module of the five modules connected in the string. Table 6.7 shows the irradiance for the different patterns. The I-V curves of these patterns can be found in Appendix H.

Figure 6.36 shows the P-V curve of pattern1.
 As can be seen from Figure 6.36, the GMPP occurs at 746.8 watts at a voltage of 122.2V. The corresponding regulated PV voltages can be found in Appendix I.

Table 6.7:Irradiences exposed to each module (W/m^2)

Pattern 1	Pattern 2	Pattern 3	Pattern 4	Pattern 5
1000	400	300	300	900
1000	400	300	1000	400
800	600	600	800	800
1200	800	800	1200	1000
500	1000	1000	900	700

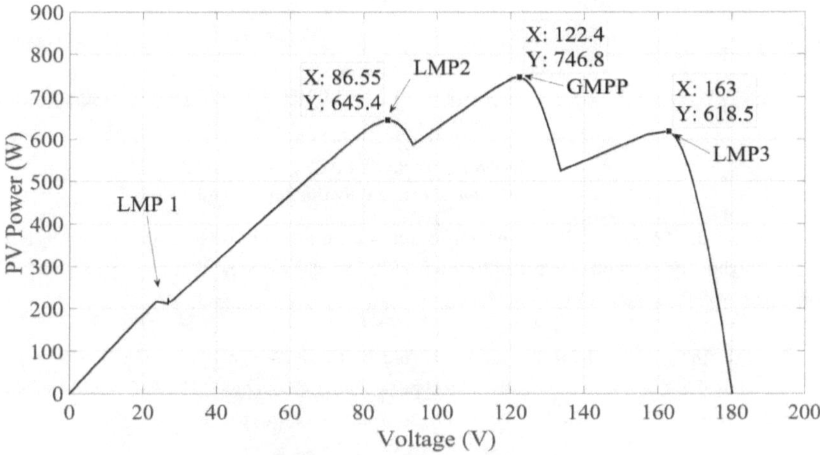

Figure 6.36 P-V curve of Pattern 1

Figure 6.37 shows the results of the power tracked by the PSO and FA

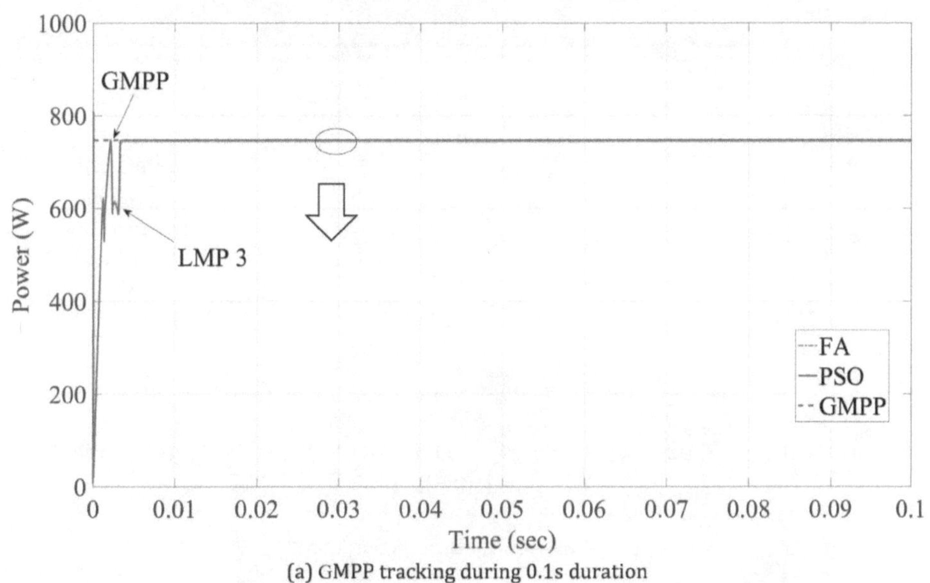

(a) GMPP tracking during 0.1s duration

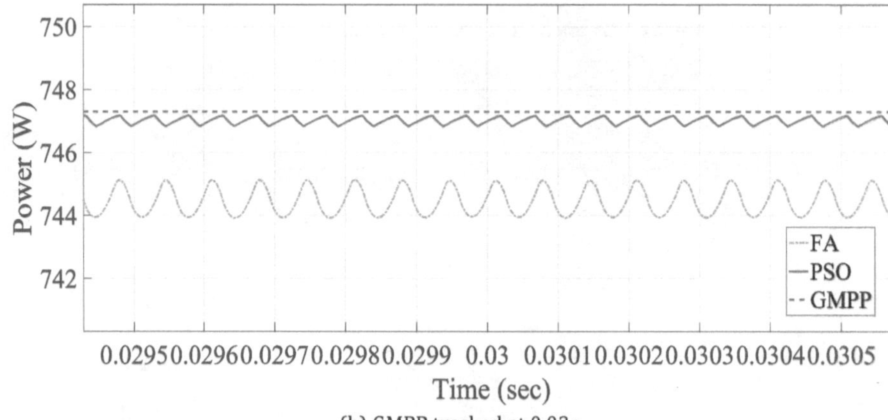

(b) GMPP tracked at 0.03s

Figure 6.37 :PV power tracked by the PSO and FA

It can be seen that the GMPP and the LMP3 are found but the algorithms were able to distinguish which MPP gives the highest power and converged to the GMPP. The PSO converged at the GMPP and the FA was very close to the GMPP as seen in Figure 6.37 (b).

Figure 6.38 shows the results of the PV power tracked by the PnO algorithm and the different starting points. From the profile there are four peaks, 3 local peaks and 1 global peak. The PnO algorithm was made to start from 0V, 60V, 100V and 140V and a constant step size of 2V was chosen.

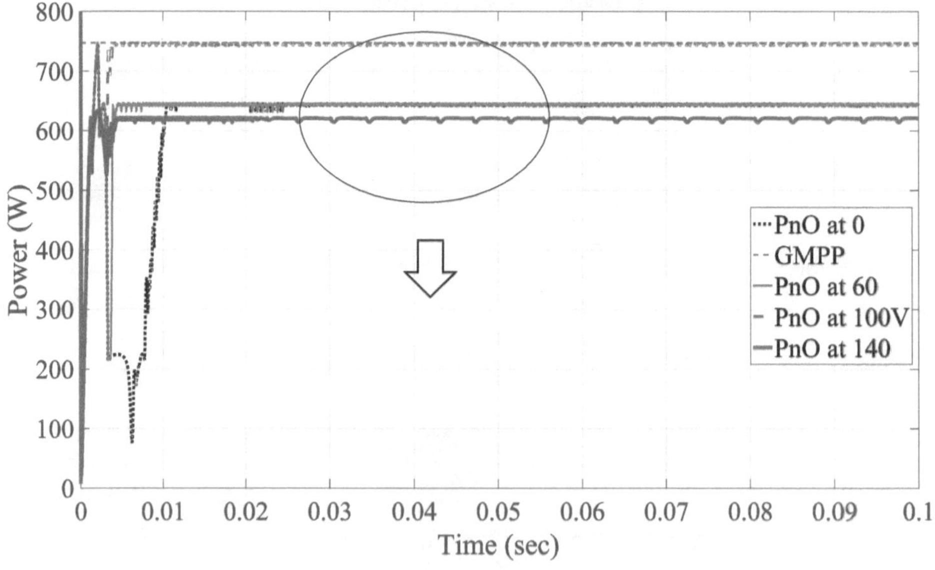

(a) MPPT for the 0.1s duration

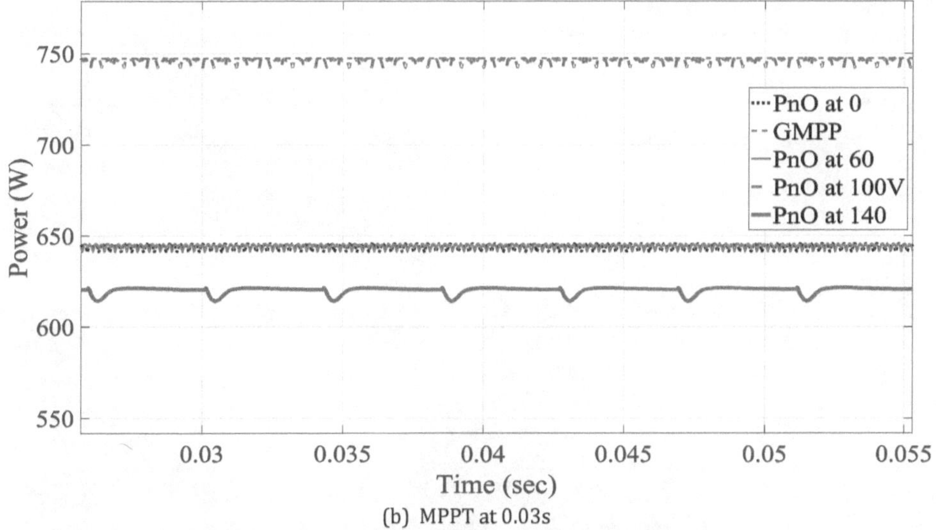

(b) MPPT at 0.03s

Figure 6.38 MPPT of the PV power for pattern1 by the PnO

It can be seen from Figure 6.38 (a) that starting the PnO from different stating points yields different results. If the algorithm is started from 0V and from 60 V, it converges to LMP2 (645.4W). If it is started from 100V the algorithm converged to the global MPP, (746.8W) and if it started at 140V the algorithm converged to LMP3 (618.5W). As expected, as soon as the PnO algorithm detects a peak it converges to it.

Figure 6.39 shows the P-V curve of pattern 2. Figure 6.40 shows the results for the FA and the PSO converging close to the GMPP.

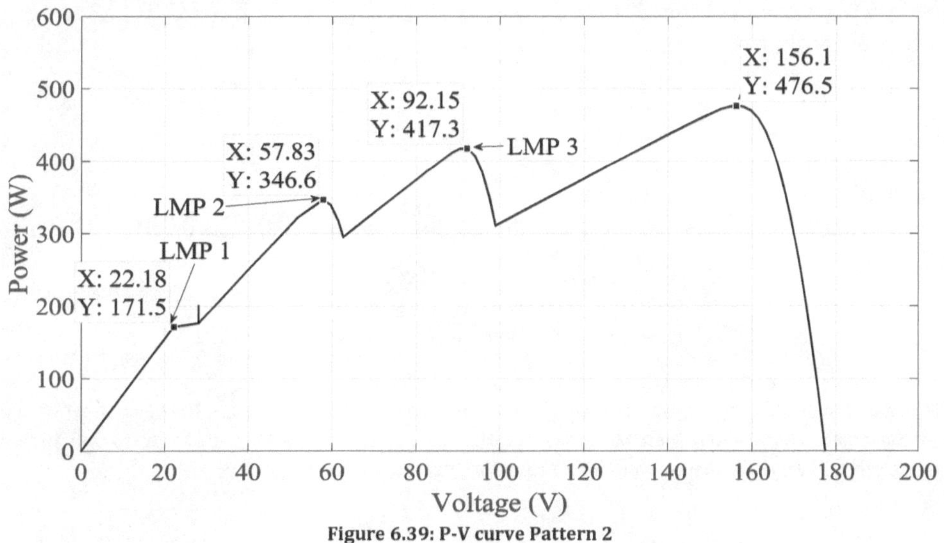

Figure 6.39: P-V curve Pattern 2

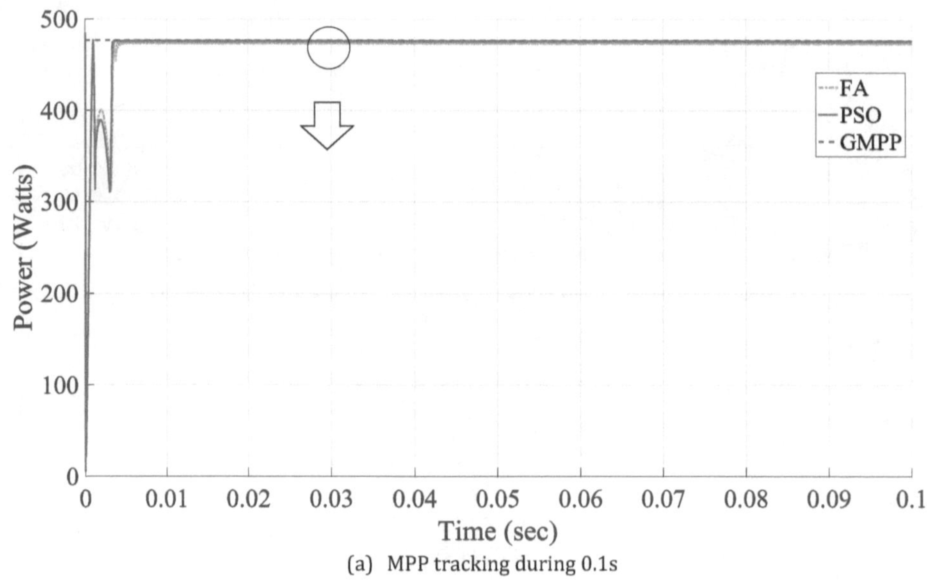

(a) MPP tracking during 0.1s

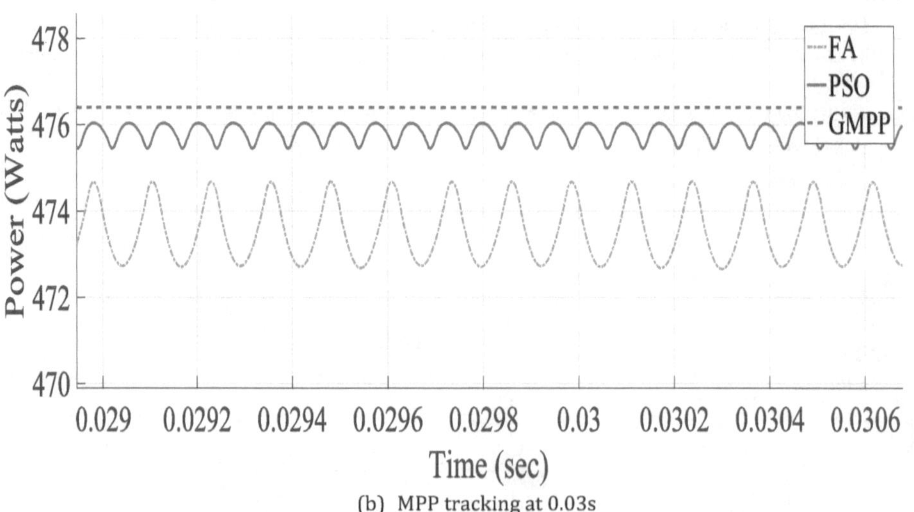

(b) MPP tracking at 0.03s

Figure 6.40: PSO and FA converging close to GMPP

Figure 6.41 shows the exploration process of the FA in finding the best Vref value. The value at which the FA converged is 154V whilst the GMPP convergence occurs at 156.1V.The PI controller takes time to find the reference as the parameters were found based on trial and error.

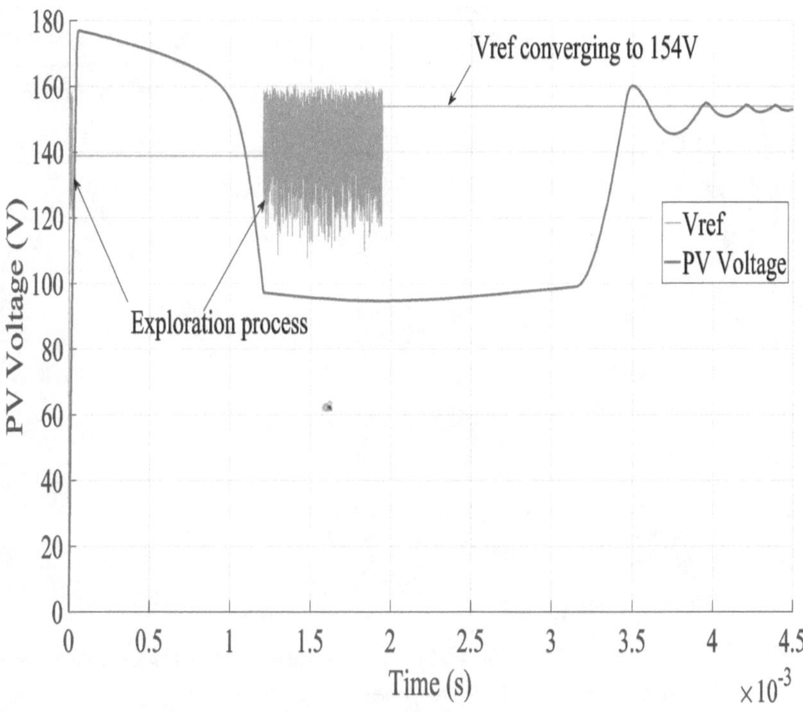

Figure 6.41 :Exploration process of Fireflies to find best Vref

Figure 6.40 shows that the PSO and FA were able to find the global maximum point of pattern 2. There is a little amount of ringing that occurs at steady state as seen in Figure 6.40 (b).

For pattern 2 the PnO was made to start from 0V, 40V, 70V and 120V. Figure 6.42 shows the PnO algorithm converging to the GMPP despite the different starting points of the algorithm. It can be seen that the one that starts at 120V converged quicker then followed by 70V, 40V and then 0V. The results are unexpected as the one that starts from 0V and 40V skipped all the LMP of pattern 2, the peaks that were skipped can be observed in Figure 6.42 especially for the PnO that starts at 0V and eventually converged to the GMP and the one that starts at 70V and skipped LMP 3. The step size of 2V was kept the same for all simulations. Figure 6.42 also shows the ringing at steady state this is expected because of the constant step size

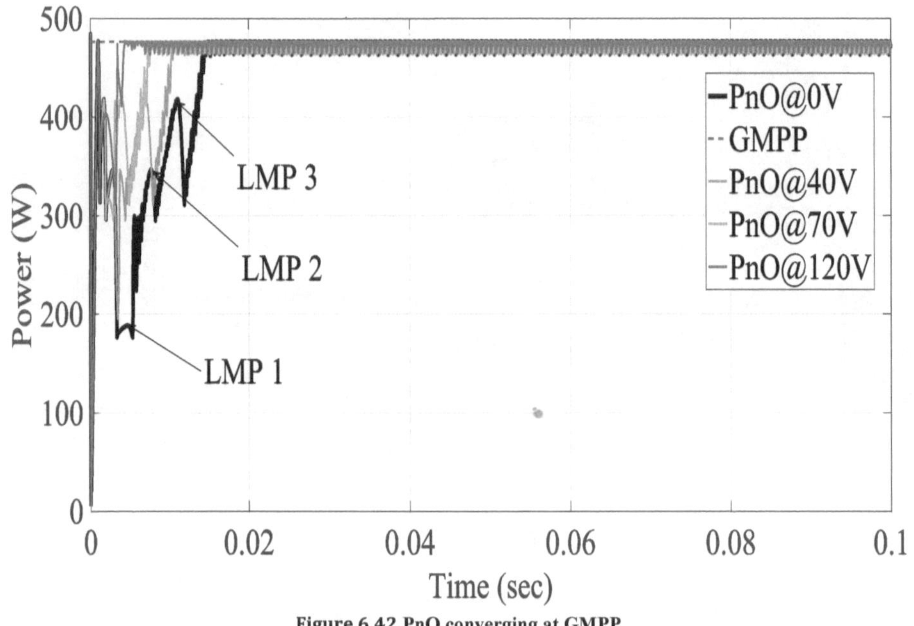

Figure 6.42 PnO converging at GMPP

Figure 6.43 shows the partial shading pattern 3. Figure 6.44 shows the PSO and FA converging close to the GMPP. The algorithm finds LMP 3 and the GMPP but decides to select the GMPP.

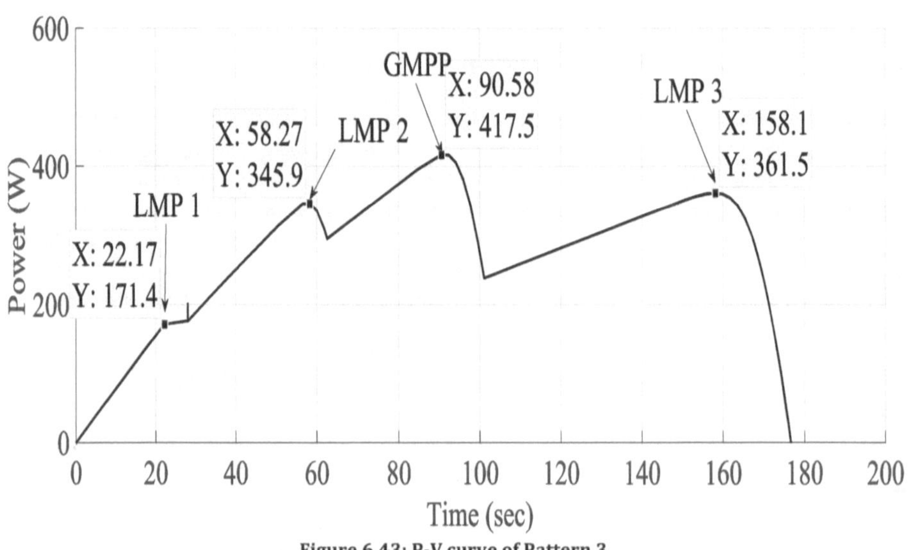

Figure 6.43: P-V curve of Pattern 3

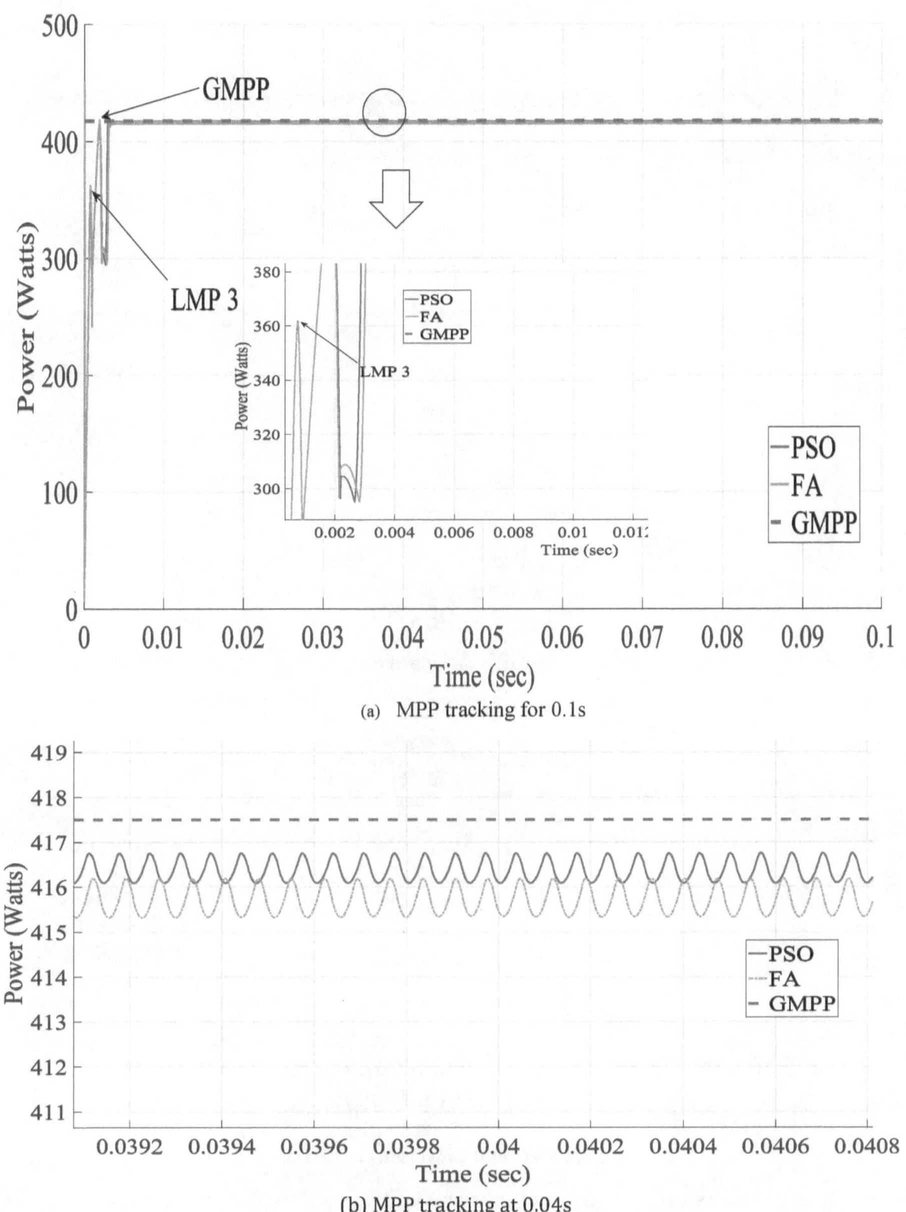

(a)   MPP tracking for 0.1s

(b) MPP tracking at 0.04s

Figure 6.44 PSO and FA converging close to the GMPP

For pattern 3, the PnO was made to start from 0V, 40V, 70V and 120V.Figure 6.45 shows the results

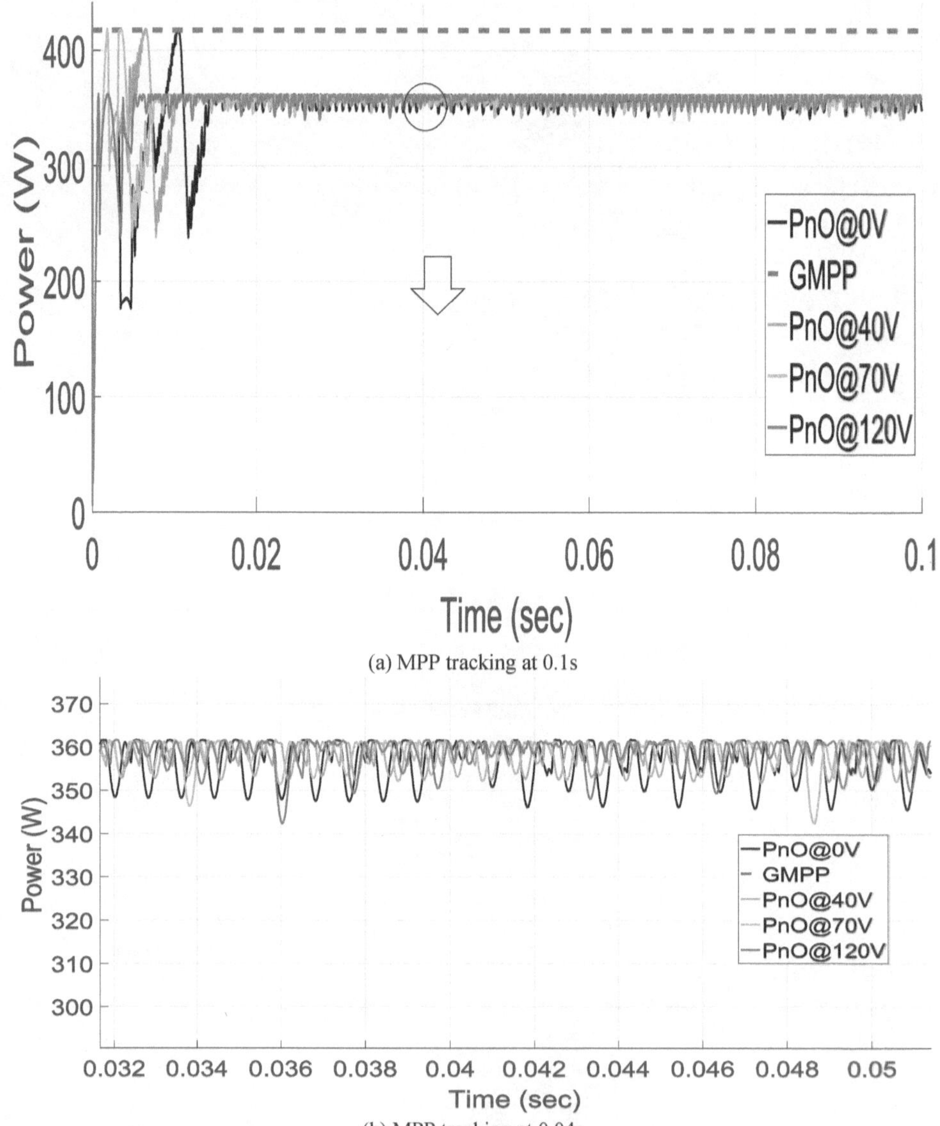

(a) MPP tracking at 0.1s

(b) MPP tracking at 0.04s

Figure 6.45: PnO converging at LMP 3

The PnO converged at the LMP 3 which is not the global best. Again when it started at 0V, 40V and 70 V it skipped the closest peak to where it starts. The PnO algorithm skipped all the peaks including the GMP to converge to the last peak. The one that started at 120V quickly converged to LMP 3 as can be seen in Figure 6.45. This clearly illustrate the lack of intelligence of this algorithm

Figure 6.46 shows pattern 4. It has four LMP and one GMP at 741.1 W.

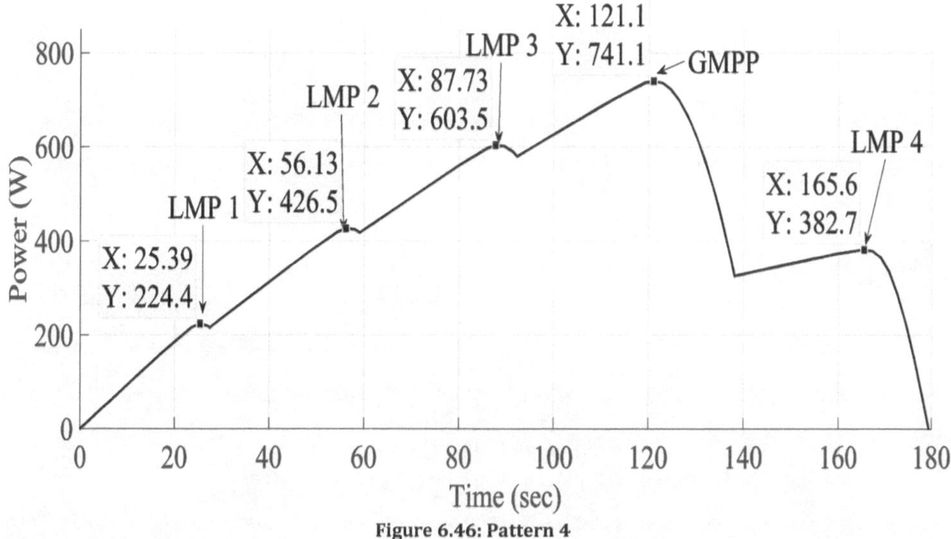

Figure 6.46: Pattern 4

Figure 6.47 shows the PSO and FA selecting the GMPP for pattern 4. The algorithms found LMP 4, LMP 3 and the GMP.

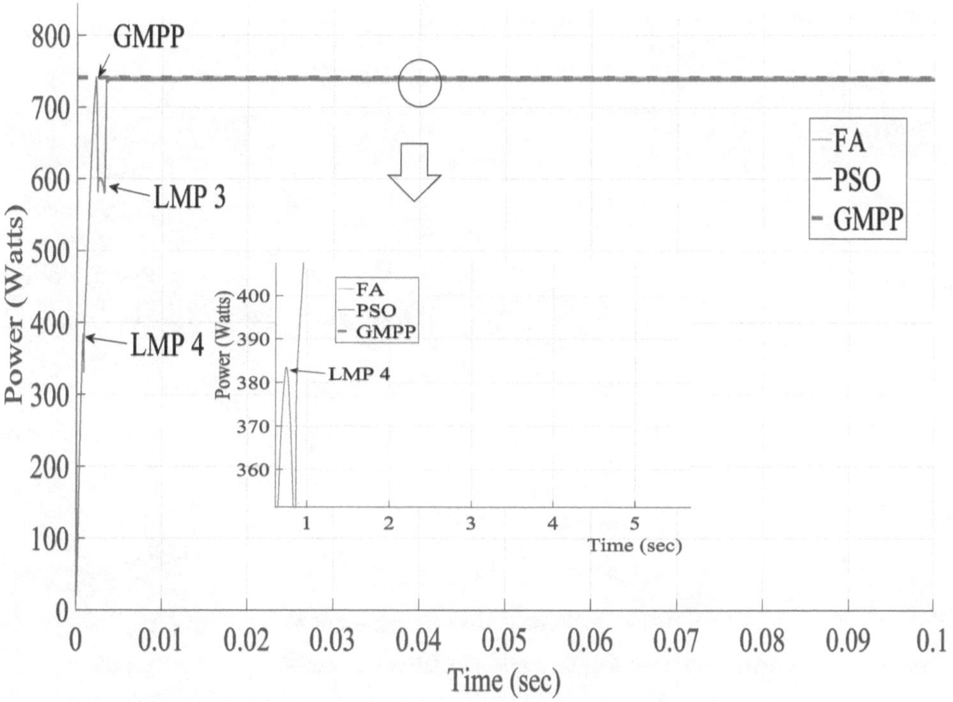

(a) MPP tracking for 0.1s

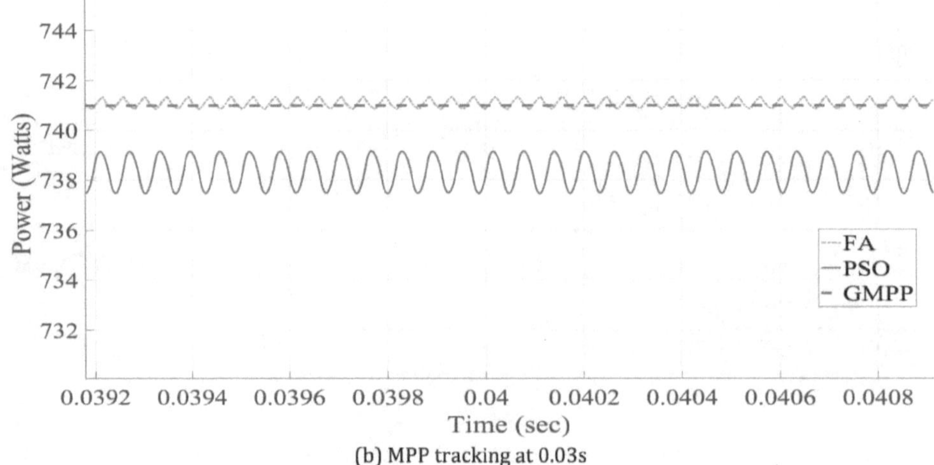

(b) MPP tracking at 0.03s

Figure 6.47 :PSO and FA converging close to GMPP

The PnO algorithm was made to start at 0V, 40V, 100V and 145V. As can be seen from Figure 6.48, the PnO that starts at 0V skipped all the peaks and converged at LMP 4. The PnO that starts at 40V also did the same thing. Even though the PnO was started at 100V right next to the GMPP, it skipped it to converge to LMP 4. As expected, the PnO that started at 145V quickly converged to LMP 4.

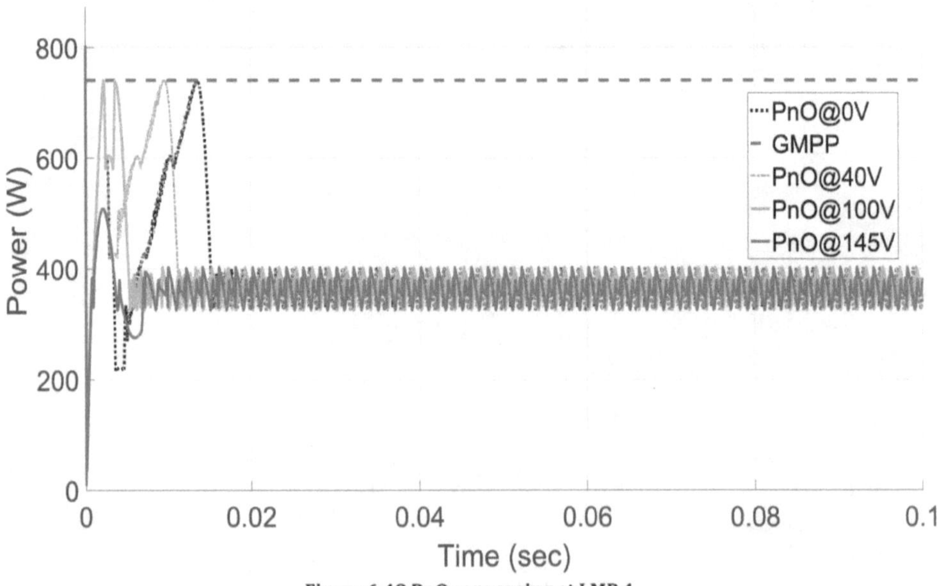

Figure 6.48 PnO converging at LMP 4

Figure 6.49 shows pattern 5 it has four LMP and one GMP at 651.4 W.

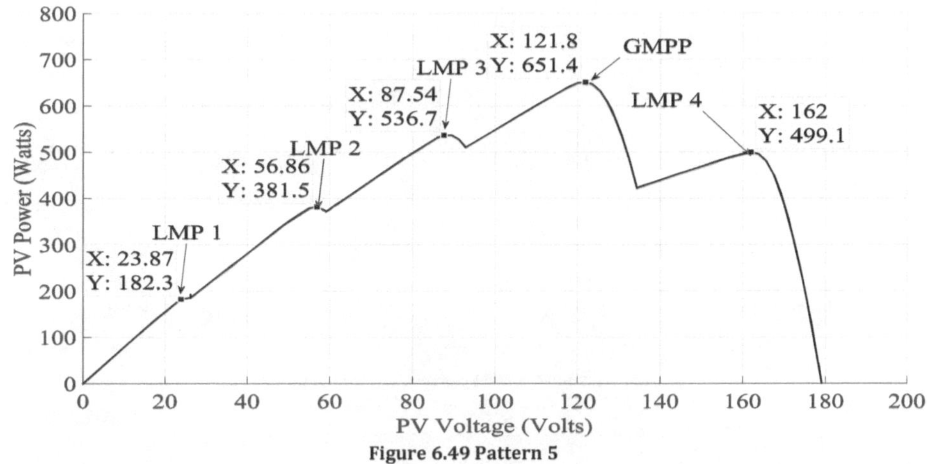

Figure 6.49 Pattern 5

The PSO and FA were able to distinguish from LMP4, LMP 3 and GMP for pattern 5 as can be seen in Figure 6.50.

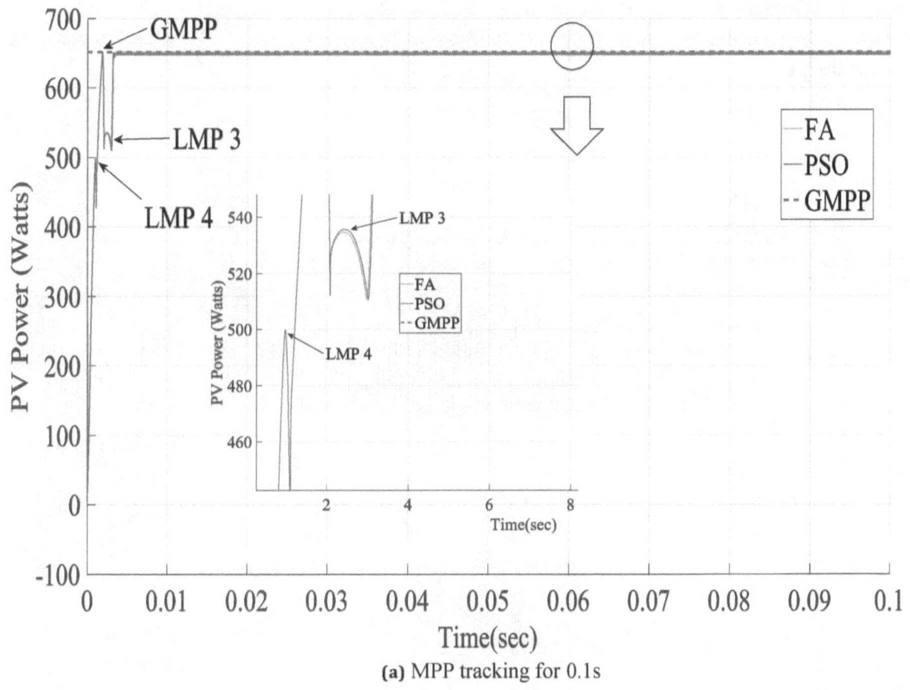

(a) MPP tracking for 0.1s

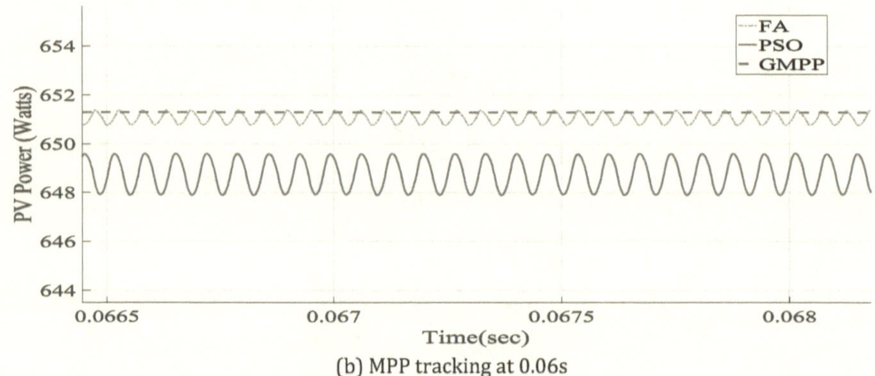

(b) MPP tracking at 0.06s

Figure 6.50 PSO and FA converging close to GMPP

The PnO algorithm was made to start from 0V, 70V, 100V and 140V. As can be seen in Figure 6.51, the PnO that starts from 0V converged at LMP2, and the one that started from 70V skipped LMP 3 and converged to the GMPP. As expected the PnO that starts from 100V converged to the peak that is right next to where it starts which is the GMPP and the PnO that started from 140V converged to LMP 4, the peak that's close to where it started.

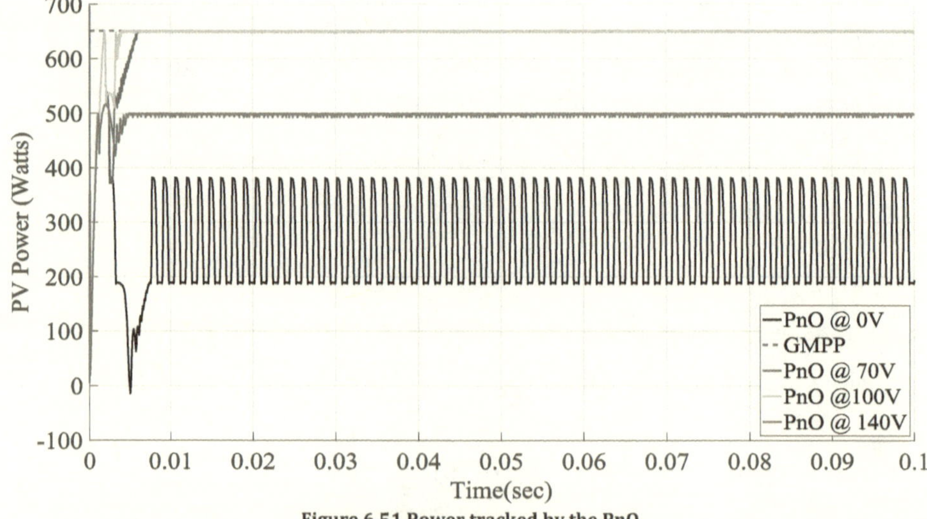

Figure 6.51 Power tracked by the PnO

Table 6.8 shows the efficiency of the algorithms. Due to the ringing of the PnO algorithm the average between the highest and lowest was taken for the values shown in Table 6.8. For pattern one, the PSO had the highest tracking efficiency followed by the FA, depending on where the algorithm was initiated for the PnO, it had various efficiencies. It can be deducted that a tracking efficiency decrease of 13.7% obtained if the PnO algorithm is initiated from 0V or 60V if compared to the PSO and if initiated at 140V a decreases of efficiency of 16.9% can be calculated.

For pattern two, the PnO tracked the GMPP despite the different starting points, the efficiencies for all algorithms are relatively the same.

For pattern 3, the PnO tracked the LMP 3 of 361W. A decrease of tracking of 14.7% can be deduced compared to the PSO.

For pattern 4, the PnO tracked LMP 4 of 370W. A decrease of tracking of 50% can be deduced compared to the FA.

For pattern 5, the FA had the highest tracking efficiency followed by the PSO, depending on where the algorithm was initiated and for the PnO it had various efficiencies. It can be deducted that a tracking efficiency decrease of 55.3% obtained if the PnO algorithm is initiated from 0V. However, if initiated at 70V a tracking efficiency decrease of 0.5% is experienced. When compared to the FA and if the PnO is initiated at 140V a decreases of efficiency of 23.5% is calculated.

Table 6.8 :Comparison between different studied MPPT techniques under PSC

Shading pattern	Technique	Power, tracked (W)	Global power (W)	Tracking efficiency %
Pattern 1	PSO	747	747.2	99.9
	FA	745		99.7
	PnO at 0V	644		86.2
	PnO at 60V	644		86.2
	PnO at 100V	744		99.6
	PnO at 140V	620		83.0
Pattern 2	PSO	476	476.1	99.9
	FA	474		99.6
	PnO	470		98.7
Pattern 3	PSO	416.5	417.5	99.7
	FA	415.7		99.5
	PnO	355		85.0
Pattern 4	PSO	739	741.1	99.7
	FA	741		99.9
	PnO	370		49.9
Pattern 5	PSO	649	651.3	99.6
	FA	650		99.8
	PnO at 0V	290		44.5
	PnO at 70V	647		99.3
	PnO at100V	647		99.3
	PnO at 140V	497		76.3

6.5 Summary

The results showed that the CI based algorithms significantly improved the PV power extraction in terms of reducing oscillations about the steady state and converging closer to the theoretical PV power compared with the PnO and IC. It was observed that considerable ringing at steady state of the PnO and IC will always exist because of the constant step size in the algorithms. The results showed that the CI and conventional algorithms had relatively close efficiencies which were all high around 99% under these static irradiance conditions and about 65% of power is lost if a direct coupled resistive load is connected.

Second, a rapid varying solar irradiance test profile was used to further investigate the CI algorithms compared to the conventional algorithms. The CI algorithms were able to extract the maximum power under rapid varying irradiance without any complications. The level of ringing at steady state was very minimum especially with the PSO and FA. They all followed closely the theoretical PV power. The PnO and IC had some problems under these conditions. The PnO was getting lost under rapid increasing irradiance and the IC was ringing more under low levels of irradiance. The results further proved that the CI algorithms are better than the conventional algorithms.

Third, five partial shading conditions were used to test the global maximum power point tracking of the PSO and FA. The PnO was used to show how it fails to track the GMPP under these conditions. The GMPP was occurring at different voltage values in the different patterns. The PSO and FA were able to distinguish the GMPP from the multiple LMPP in all partial shaded patterns. The PnO was shown to be unreliable in tracking the GMPP. This was expected as it lacks intelligence to know which peak gives the best power. The results also show that starting the PnO algorithms at different points can yield different results. Due to the ability of the PSO and FA for global searching in a multimodal environment they tracked the GMPP all the time for each partial shaded pattern.

The results obtained in the book are similar to the results mentioned in literature. That is, the level of ringing at steady state using PSO and FA is very minimum compared to the PnO and IC. The results of the PSO and FA algorithms searching globally to find the best variable to produce the optimum output agree with literature.

Chapter 7 Conclusion and Recommendations

7.1 Conclusion

It can be seen from the results that the PnO algorithm is unpredictable in tracking the GMP under partial shading conditions. It can directly converge to the LMP right next to where it starts or it might skip the LMP and converge to another LMP or it might converge to the GMP.

MPPT algorithms were successfully modelled and implemented. The challenge with the classical conventional algorithms is that the ringing at steady state is unavoidable. Under rapid varying irradiance, the PnO gets lost when the irradiance keeps on increasing, this is due to the algorithm not being able to distinguish whether the change in power is from perturbation or a change in irradiance. The IC is unstable at low irradiance as the ringing increased under these circumstances.

The stochastic based computational algorithms have intelligence and are able to find which Power is the brightest (firefly algorithm) so they are able to track the GMPP for every pattern. From the population initiated and then randomized in the search space, the best voltage value that gives the best power is able to be found. The level of ringing that occurs at steady state is very minimum.

The use of stochastic based computational intelligent algorithms will play an important role in PV systems as partial shaded conditions are very common. The level of power loss due to ringing once the algorithm converges is minimum in stochastic algorithms than in deterministic algorithms.

Looking at the efficiency of the PSO and FA, they performed relatively the same. This was expected as they are both stochastic based optimisation algorithms which use relatively the same principle in tracking GMPP in a multimodal environment. What differs in these algorithms is the number of tuning parameters. Ideally, less parameter to tune would be required and this would mean simplicity and faster converge. The PSO has three parameters to tune (w, $c1$ and $c2$) and the FA has two (α and γ). FA having only two variables to tune means it is simpler and quicker to converge compared to PSO which has three variables to tune.

The drawbacks of the PSO and FA are that it requires a great deal of computing compared to PnO and their convergence relies heavily on their parameters. The fuzzy logic drawbacks are that the membership functions have to be tuned according to a rule base table to accomplish satisfactory results which can be a challenging task. All the computational intelligence algorithms were seen to be superior in performance compared to the conventional algorithms in all aspects.

Computational intelligence can be successfully used for partial shaded conditions as shown in this book. These methods are less complex than hardware fixture methods that exist.

7.2 Recommendation

The convergence of stochastic based computational intelligent algorithms relies heavily on the parameters of the algorithms, future work can be done to see the effect of varying these parameters to improve convergence in MPPT in PV systems.

To verify the simulations done in this book a practical hardware tests at a reasonable PV power level and under real world atmospheric conditions can be implemented. The MPPT techniques developed in this book will hopefully incentivise building a prototype model to investigate and compare experimentally.

In boost controller design, metaheuristic algorithms can be used to find the best PI controller parameters that will best fit controller design specifications, i.e. minimum overshoot, minimum settling time, minimum rise time etc. The multi objectives can be made into a mathematical function then minimised.

List of References

[1] G.M.Masters, "Renewable and efficient electric power systems," in *Wiley-Interscience*, 2005.

[2] F. A. O. Aashoor, "Maximum power point tracking techniques for photovoltaic water pumping system," p. 217-223, 2015.

[3] B. Tull, "Photovoltaic Cells: Science and Materials," 2004. [Online]. Available: http://www.molchem. science.ru.nl/rowan/Coll/caput-college/PhotovoltaicCells.pdf. [Accessed monday march 2017].

[4] M. G. Villalva, "Comprehensive Approach to Modeling and Simulation of Photovoltaic Arrays,," *Power Electronics, IEEE Transactions,* vol. 24, pp. 1198-1208, 2009.

[5] T. Huan-Liang, "Insolation-oriented model of photovoltaic module using Matlab/Simulink," *Solar Energy,* vol. 84, pp. 1318-1326, 2010.

[6] G. Walker, "Evaluating MPPT converter topologies using a MATLAB PV model,," *Journal of Electrical & Electronics Engineering, Australia,* vol. 21`, pp. 49-56, 2001.

[7] E. M. Natsheh, "Modeling and control for smart grid integration of solar/wind energy conversion system," in Innovative Smart Grid Technologies („," in *IEEE PES International Conference and Exhibition,* ISGT Europe), 2011.

[8] S. R. Wenham, "S. R. Wenham," Earthscan/James & James, 2007.

[9] R. Y. F. Habbati Bellia, "A detailed modeling of photovoltaic module," *NRIAG Journal of Astronomy and Geophysics,* vol. 3, pp. 53–61, 2014.

[10] R. C. Neville, "Solar energy conversion: the solar cell," *Elsevier Science,* 1995.

[11] Krismadinata, "Photovoltaic Module Modeling using Simulink/Matlab," *Procedia Environmental Sciences,* vol. 17, pp. 537-546, 2013.

[12] C. M. F. Santos, *Optimised Photovoltaic Solar Charger With Voltage Maximum Power Point Tracking,* Master book, Universidade Técnica de Lisbo, 2008.

[13] Y. T. Tan, *Impact on the power system with a large penetration of photovoltaic generation,* PhD, Manchester, 2004.

[14] T. Kerekes, E. Koutroulis, D. Sera, R. Teodorescu, and M. Katsanevakis, ""An optimization method for designing large PV plants," *IEEE,* vol. 3, no. 2, pp. 814–822, 2013.

[15] Lumby, A. Miller and B., "Utility Scale Solar Power Plants: A Guide for Developers and Investors.," Washington, DC, USA: World Bank, 2012.

[16] S. Titri, C. Larbesb, K.Youcef Toumic, and K.Benatchbada, ""A new MPPT controller based on the Ant colony optimization algorithm for Photovoltaic systems under partial shading conditions," *Applied Soft Computing,* vol. 58, pp. 465–479, 2017.

[17] A.Benyoucefa, A. Chouder , K. Karaa, and S. Silvestrec, "Artificial bee colony based algorithm for maximum power point tracking (MPPT) for PV systems operating under partial shaded conditions," *Applied Soft Computing,* vol. 32, pp. 38–48, 2015.

[18] C. F. Fang ZHANG Jian LI, "In-depth Investigation of Effects of Partial Shading on PV Array Characteristics," *IEEE* Power Engineering and Automation Conference, 2012.

[19] G. Velasco-Quesada, F. Guinjoan-Gispert, R. Piqué-López,M. Román-Lumbreras, and A. Conesa-Roca, "Electrical PV array reconfiguration strategy for energy extraction improvement in grid connected PV systems," *IEEE Trans. Ind. Electron,* vol. 56, no. 11, pp. 4319–4331,, 2009.

[20] D. Nguyen and B. Lehman, "An adaptive solar photovoltaic array using model-

reconfiguration algorithm," *IEEE Trans. Ind. Electron,* vol. 55, no. 7, pp. 2644–2654, 2008.

[21] N. Femia, G. Lisi, G. Petrone, G. Spagnuolo, and M. Vitelli, "Distributed maximum power point tracking of photovoltaic arrays: Novel approach and system analysis," *IEEE Trans. Ind. Electron,* vol. 55, no. 7, pp. 2610–2621, 2008.

[22] Pilawa-Podgurski, R. Bell and R. C. N., "Decoupled and distributed maximum power point tracking of series-connected photovoltaic submodules using differential power processing," *IEEE J. Emerg. Sel. Topics Power Electron,* vol. 3, no. 4, pp. 881–891, 2015.

[23] L. F. L. Villa, T.-P. Ho, J.-C. Crebier, and B. Raison, ""A power electronics equalizer application for partially shaded photovoltaic modules,," *IEEE Trans.,* vol. 60, no. 3, pp. 1179–1190,, 2013.

[24] M. Uno and A. Kukita, "Single-switch voltage equalizer using multistacked buck–boost converters for partially shaded photovoltaic modules," *IEEE Trans. on Power Electron,* vol. 30, no. 6, p. 3091–3105, 2015.

[25] Blaabjerg, E. Koutroulis and F., "A new technique for tracking the globalmaximum power point of PV arrays operating under partial-shading conditions," *IEEE J. Photovolt,* vol. 2, no. 2, pp. 184–190, 2012.

[26] N. Mohan and T. M. Undeland, Power electronics: converters, applications, and design, Wiley-India, 2007.

[27] M. H. Rashid, Power Electronics - Circuits, Devices, and Applications 3rd Edition Pearson Education, Academic Pr, 2004.

[28] Xin-SheYang, Engineering Optimization An Introduction with Metaheuristic Applications, Contemporary Physics,2012.

[29] X.-s. Yang, "Nature-Inspired Metaheuristic Algorithms," in *Engineering Optimization An Introduction with Metaheuristic Applications*, UK, University of Cambridge, 2010 .

[30] S. S. Arora, "The Firefly Optimization Algorithm:Convergence Analysis and Parameter Selection",," *International Journal of Computer Applications,* vol. 69, no. 3, pp. 0975 – 8887, 2013 .

[31] A. O. Omole, "Analysis, modeling and simulation of optimal power tracking of multiple-modules of paralleled solar cell systems," Master of Science book, The Florida State University College of Engineering, 2006.

[32] K. H. Hussein, "Maximum photovoltaic power tracking: an algorithm for rapidly changing atmospheric conditions," *Generation, Transmission and Distribution, IEE Proceedings,* vol. 142, pp. 59-64, 1995.

[33] E. Koutroulis, "Development of a microcontroller-based, photovoltaic maximum power point tracking control system," *IEEE Transactions on Power Electronics,* vol. 16, pp. 46-54, 2001.

[34] L. Fangrui, "Comparison of PnO and hill climbing MPPT methods for grid-connected PV converte," in *Industrial Electronics and Applications, 2008. ICIEA 2008. 3rd IEEE Conference on,* 2008.

[35] T. Noguchi, "Short-current pulse-based maximum-power-point tracking method for multiple photovoltaic-and-converter module system," *IEEE Transactions on Industrial Electronics, ,* vol. 49, pp. 217-223, 2002.

[36] J. Ahmad, "A fractional open circuit voltage based maximum power point tracker for photovoltaic arrays," *Software Technology and Engineering (ICSTE), 2010 2nd International Conference,* vol. 1, pp. V1-247-V1-250., 2010.

[37] Chapman, T. Esram and P. L., "Comparison of Photovoltaic Array Maximum Power Point Tracking Techniques," *Energy Conversion, IEEE Transactions,* vol. 22, pp. 439-449, 2007.

[38] V. Salas, "Review of the maximum power point tracking algorithms for stand-alone photovoltaic systems," *Solar Energy Materials and Solar Cells,* vol. 90, pp. 1555-1578,, 2006.

[39] T. Hiyama, "Identification of optimal operating point of PV modules using neural network for real time maximum power tracking control," *Energy Conversion, IEEE Transactions on,* vol. 10, pp. 360-367, 1995.

[40] M. Kadir, "A Novel MPPT Method for PV Arrays Based on Modified Bat Algorithm with Partial Shading Capability," *International Journal of Computer Science and Network Security,* vol. 17, no. 2, 2017.

[41] Satyajit Mohanty,Bidyadhar Subudhi,Pravat Kumar Ray, "A New MPPT Design Using Grey Wolf Optimization Technique for Photovoltaic System Under Partial shading," *IEEE Transactions on Sustainable Energy,* vol. 7, no. 1, pp. 220-228,2016.

[42] Jiying Shi, Wen Zhang∗, Yongge Zhang, Fei Xue, Ting YangSchool, "MPPT for PV systems based on a dormant PSO algorithm," *Electric Power Systems Research,* vol. 123, pp. 100-107, 2015.

[43] D. F. Teshome, C. H. Lee, Y. W. Lin, and K. L. Lian, "A Modified Firefly Algorithm for Photovoltaic Maximum Power Point Tracking Control Under Partial Shading," *Ieee journal of emerging and selected topics in Power Electronics,* vol. 5, no. 2, 2017.

[44] N. Mohan, Power Electronics,Department of Electrical Engineering University of Minnesota,1995.

[45] X. Weidong and W. G. Dunford, "A modified adaptive hill climbing MPPT method for photovoltaic power systems," *Power Electronics Specialists Conference,,* vol. 3, pp. 1957-1963, 2004.

[46] D. P. Hohm and M. E. Ropp,, "Comparative study of maximum power point tracking algorithms using an experimental, programmable, maximum power point tracking test bed,," *Photovoltaic Specialists Conference,,* pp. 1699-1702., 2000.

[47] T. Tafticht, "A new MPPT method for photovoltaic systems used for hydrogen production," COMPEL," *The International Journal for Computation and Mathematics in Electrical and Electronic Engineering,* vol. 26, pp. 62-74, 2007.

[48] T. Tafticht, "An improved maximum power point tracking method for photovoltaic systems,," *Renewable Energy,* vol. 33, pp. 1508-1516, 2008.

[49] A. Oi, "Design and Simulation of Photovoltaic Water Pumping System," Faculty of California Polytechnic State University, 2005.

[50] N. Pongratananukul, "Analysis and Simulation Tools for Solar Array Power Systems," University of Central Florida, 2005.

[51] T. E. Persen, "FPGA-Based Design of a Maximum-Power-Point-Tracking System for Space Application," University of Florida, 2004.

[52] Salas V , Olias E, Barrado A, Lazaro A. "Review of the maximum power point tracking algorithms for," *Solar Energy Materials & Solar Cells,* vol. 90, pp. 1555—1578, 2006.

[53] Kermadi M, El Madjid Berkouk, "Artificial intelligence-based maximum power point tracking controllers for Photovoltaic systems: Comparative study," *Renewable and Sustainable Energy Reviews,* vol. 69, pp. 369–386, 2017.

[54] Trishan Esram, and Patrick L. Chapman, "Comparison of Photovoltaic Array Maximum Power Point Tracking Techniques," *IEEE transactions on energy conversion,* 2006.

[55] Xiao W, "A Modified Adaptive Hill Climbing Maximum Power Point Tracking (MPPT) Control Method For Photovoltaic Power Systems," The University of British Columbia, 2003.

[56] Ting-Chung Yu ,Yu-Cheng Lin, "A Study on Maximum Power Point Tracking Algorithms for Photovoltaic," Department of Electrical Engineering Lunghwa University of Science and Technology, 2010.

[57] N. H. Selman, "Comparison Between Perturb & Observe, Incremental Conductance and Fuzzy Logic MPPT Techniques at Different Weather Conditions," *International Journal of Innovative*

Research in Science. Engineering and Technology, vol. 5, no. 7, 2016.

[58] Ramchandania ,V Pamarthib, K.,Chowdhury S. R., "Comparative Study of Maximum Power Point Tracking Using Linear Kalman Filter & Unscented Kalman Filter for Solar Photovoltaic array on Field Programmable Gate Array," International Journal on Smart Sensing and Intelligent Systems, vol. 5, no. 3, pp. 701-716, 2012.

[59] Chikate V, Prof. D. R. Dandekar, Prof. Mrs. Y. A. Sadawarte, "Comparison Of Perturb & Observe and Incremental Conductance Algorithms for MPPT Controller," International Journal of Advance Foundation and Research in Computer (IJAFRC), vol. 2, no. 9, pp. 2348 – 4853, 2015.

[60] R. Faranda and S. Leva, "Energy comparison of MPPT techniques for PV Systems," WSEAS transactions on power systems, vol. 3, pp. 446-455, 2008.

[61] C. Hua and C. Shen,, "Comparative study of peak power tracking techniques for solar storage system," in Applied Power Electronics Conference and Exposition, 1998.

[62] H. Knopf, "Analysis, Simulation, and evaluation of maximum power point tracking (MPPT) methods for a solar powered vehicle," Master book, Portland State University, 1999.

[63] S. Yuvarajan and X. Shanguang, ""Photo-voltaic power converter with a simple maximum-power-point-tracker," in Circuits and Systems," ISCAS '03. Proceedings of the 2003 International Symposium, vol. 3, pp. III-399-III-402, 2003.

[64] M. Abou El Ela and J. A. Roger,, "Optimization of the function of a photovoltaic array using a feedback control system,"," Solar Cells, vol. 13, pp. 107-119, 1984.

[65] F. Chekireda,*, C. Larbesa, D. Rekiouab, F. Haddadc, "Implementation of a MPPT fuzzy controller for," Energy Procedia, vol. 6, pp. 541–549, 2011.

[66] H.E.A. Ibrahim,M Ibrahim, "Comparison Between Fuzzy and PnO Control for MPPT for," Journal of Energy Technologies and Policy, vol. 6, No.6, 2012.

[67] Mukesh Kumar, Dr S.R.Kapoor, Rajkumar Nagar, Amit Verma, "Comparison between IC and Fuzzy Logic MPPT Algorithm Based Solar PV System using Boost Converter," International Journal of Advanced Research in Electrical,Electronics and Instrumentation Engineering, vol. 4, no. 6,pp. 2708, 2015.

[68] Ying Bai and Dali Wang, "Fundamentals of Fuzzy Logic Control – Fuzzy Sets,Fuzzy Rules and Defuzzifications," Advanced Fuzzy Logic Technologies in Industrial Applications,Advances in Indusrial Control.Springer London, pp. 17-36,2006 .

[69] Ali Bidram, Ali Davoudi, and Robert S. Balog, "Control and circuit Techniques to Mitigate Partial Shading Effects in Photovoltaic Arrays," IEEE Journal of Photovoltaics, vol. 2, no. 4, 2012.

[70] Jubaer Ahmed, Zainal Salam, "A Maximum Power Point Tracking (MPPT) for PV system using Cuckoo Search with partial shading capability," Applied Energy, vol. 119, pp. 118-130, 2014.

[71] Kok Soon Tey, Saad Mekhilef, Hong-Tzer Yang, and Ming-Kai Chuang, "A Differential Evolution Based MPPT Method for Photovoltaic Modules under Partial Shading Conditions," International Journal of Photoenergy, 2014.

[72] Gauri Mantri, N. R. Kulkarni, "Design and Optimization of PID controller using Genetic algorithm," International Journal of Research in Engineering and Technology, pp. 2321-7308, 2012.

[73] K.W.E.Cheng, Classical switched mode and resonant power converters, Hong Kong Polytechnic University, 2010. book.

[74] D. D. Corrigan, "Characterising the Response of a Closed Loop System," Electronic and Electrical Engineering, 2012 . MIT notes.

[75] Yi-Hwa Liu, , Shyh-Ching Huang, Jia-Wei Huang, and Wen-Cheng Liang, "A Particle Swarm Optimization-Based Maximum Power Point Tracking Algorithm for PV Systems Operating Under

Partially Shaded Conditions," *IEEE transactions on energy conversion,* vol. 27, no. 4, 2012.

[76] M. Clerc and J.Kennedy, "The particle swarm—Explosion, stability, and convergence in a multi-dimensional complex space," *IEEE Trans. Evol. Comput,* vol. 6, pp. 58–73, 2002.

[77] P. N. Suganthan, "Particle swarm optimizer with neighborhood operator," *IEEE Int. Congr. Evolutionary Computation,* vol. 3, pp. 1958–1962, 1999.

[78] M. Clerc, "The swarm and the queen: Toward a deterministic andadaptive particle swarm optimization," *IEEE Int. Congr.Evolutionary Computation,* vol. 3, pp. 1957, 1999.

[79] A. Ratnaweera, S. K. Halgamuge, and H. C. Watson, "Self-Organizing Hierarchical Particle Swarm Optimizer With Time-VaryingAcceleration Coefficients," *IEEE Transactions on Evol. Computation,* vol. 8, no. 3, 2004.

[80] Y. Shi and R. C. Eberhart, "Empirical study of particle swarm optimization," *Evolutionary Comput.,* pp. 1945–1950, 1999.

[81] J. Kennedy and R. Eberhart, ""Particle swarm optimization," *IEEE Int. Conf. Neural Networks,* pp. 1942–1948, 1995.

[82] K. L. Lian, J. H. Jhang, and I. S. Tian, "A Maximum Power Point Tracking Method Based on Perturb-and-Observe Combined With Particle Swarm Optimization" IEEE Journal of Photovoltaics,Vol.4,no 2,2014.

[83] ZHANG Li-ping , YU Huan-jun and, HU Shang-xu, "Optimal choice of parameters for particle swarm optimization," *Journal of Zhejiang University SCIENCE ISSN,* pp. 1009-3095, 2005.

[84] MO Yuan-bin, MA Yan-zhui, Zheng Qiao-yan, "Optimal Choice of Parameters for Firefly Algorithm," in *Fourth International Conference on Digital Manufacturing & Automation*, 2013.

[85] Yuntao Dai, Liqiang Liu, and Ying Li,, "An Intelligent Parameter Selection Method for Particle Swarm Optimization Algorithm",," in *Fourth International Joint Conference on Computational Sciences and Optimization*, 2011.

Appendices

Appendix A: Actual Data sheet of PV module parameter

1STH-215-P Solar Panel from 1Soltech

- Panels From This Manufacturer
- Add to Comparison Cart
- Request For Quote

Specifications	
Electrical Characteristics	
STC Power Rating P_{mp} (W)	215
PTC Power Rating P_{mpp} (W)	189.4
PTC/STC Power Ratio	88.1%
Open Circuit Voltage V_{oc} (V)	36.3
Short Circuit Current I_{sc} (A)	7.84
Voltage at Maximim Power V_{mp} (V)	29.0
Current at Maximim Power I_{mp} (A)	7.35
Panel Efficiency	13.7%
Fill Factor	75.5%
Power Tolerance	-3.00% ~ 3.00%
Maximum System Voltage V_{max} (V)	600
Maximum Series Fuse Rating (A)	15
Temperature Coefficients	
Temperature Coefficiency of I_{sc}	0.010 %/°C
Temperature Coefficiency of V_{oc}	-0.36 %/°C
Temperature Coefficiency of P_{mp}	-0.50 %/°C
Mechanical Characteristics	
Cell Type	Polycrystalline Cell
Cell Size(mm)	156 × 156
Cells	6 × 10
Dimensions	1626.0 × 984.0 × 46.0mm (38.0 × 64.0 × 1.8 inch)
Weight	20.0Kg (44.1 lbs)
Junction Box (Safety Rating, Bypass Diodes)	
Positive Cable (Length, Cable Cross-Section)	
Negative Cable (Length, Cable Cross-Section)	
Plug Connector (Type, Safety)	
Front Cover (Thickness, Material)	
Backsheet Cover (Color, Thickness, Material)	
Encapsulation Materials	
Frame Material	
Operation Conditions	
Nominal Operating Cell Temperature (NOCT)	47.4°C
Operating Temperature	-40.0°C to 85.0°C
Maximum Load	
Hail Storm Rating	
Fire Safety Rating	
Warranty & Certification	
Certificates	• UL 1703 • CEC California
Defects & Workmanship Warranty Period	5.0 Years
90% Power Output Warranty Period	10.0 Years
80% Power Output Warranty Period	25.0 Years
Are Warranties Insured By Third Party	False

Figure A. 1 Data sheet of actual PV module

Figure A. 2 Matlab simulink PV module block parameters

Appendix B: Mathematical modelling of the PV module

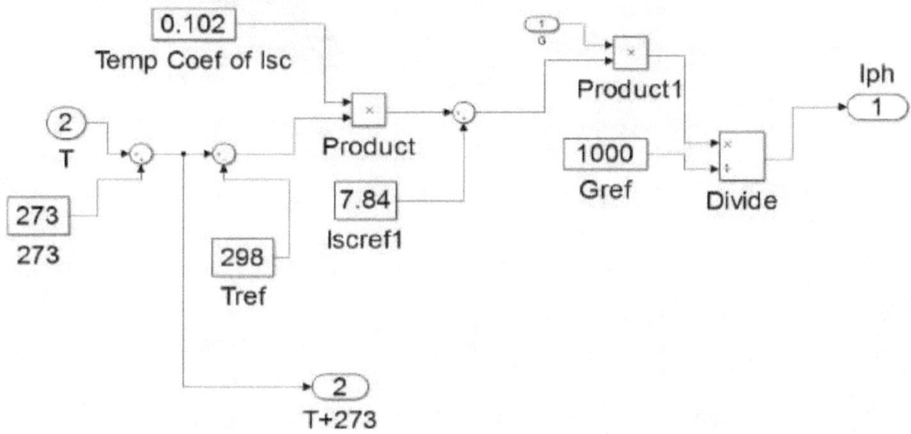

Figure B. 1 Detailed Iph implementation

$$I_{Ph} = [I_{SC} + K_1(T_C - T_{Ref})]\frac{G}{G_r} \tag{2.5}$$

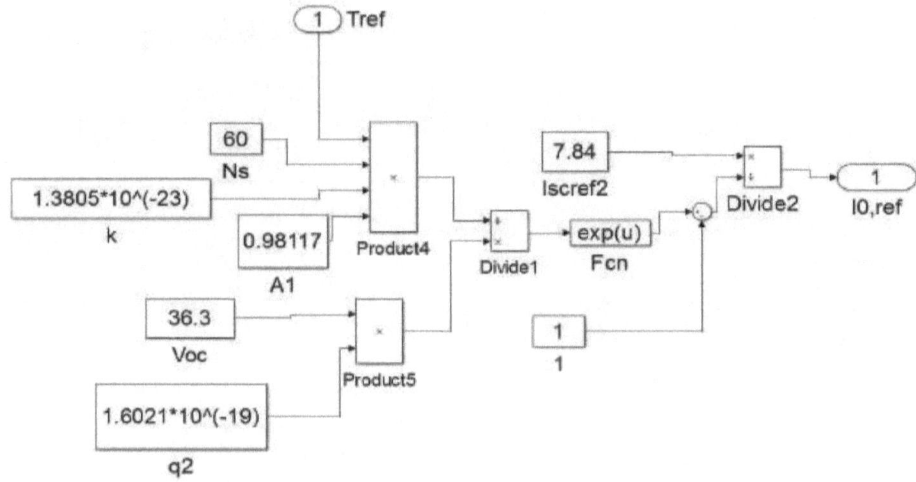

Figure B. 2 Detailed reference reverse saturation current

$$I_{RS} = \frac{I_{SC}}{\exp\left(\frac{qV_{oc}}{Ak\,T_{Ref}}\right) - 1} \tag{2.4}$$

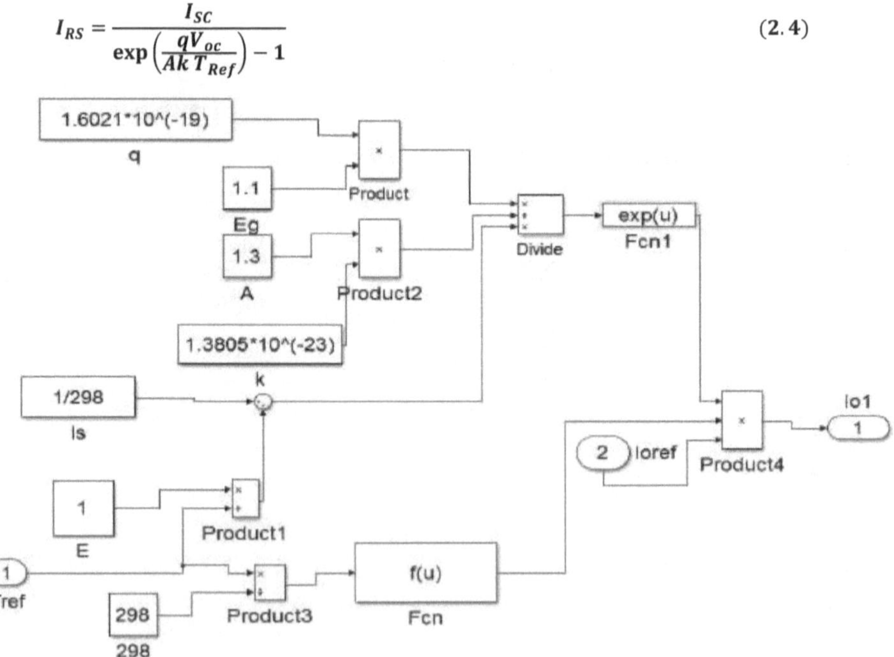

Figure B. 3 Detailed saturation current varying with temperature

$$I_S = I_{RS} \cdot (\frac{T_C}{T_{Ref}})^{(\frac{3}{A})} \cdot \exp\left[\frac{qE_{gap}}{Ak}\left(\frac{1}{T_{Ref}} - \frac{1}{T_C}\right)\right] \tag{2.3}$$

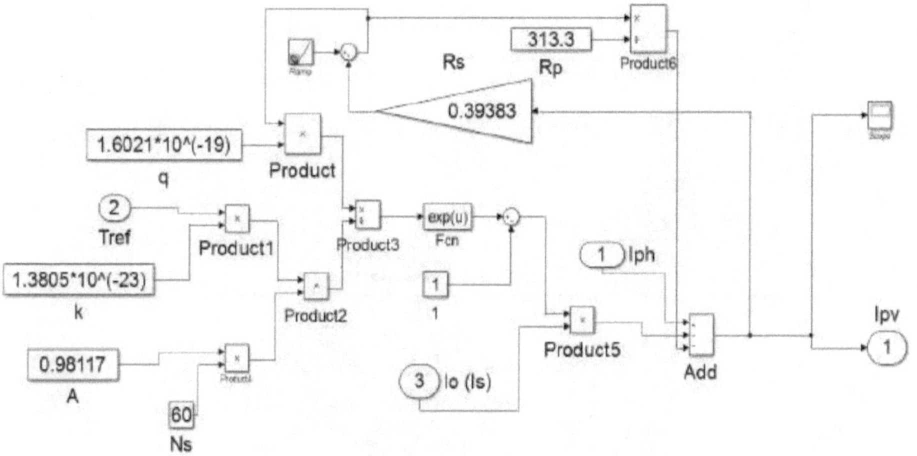

Figure B. 4 Detailed PV Current with Rs and Rsh

$$I_{PV} = I_{Ph} - I_S\left[\exp\left(\left(\frac{q(V_{PV} + R_S I_{PV})}{NsAkT_C}\right) - 1\right] - \left(\frac{V_{PV} + R_S \cdot I_{PV}}{R_{Sh}}\right) \tag{2.7}$$

Appendix C: m-file Code for boost converter small signal model and control

```
%Model of non-isolated Boost converter
Vg=145;    %Input voltage
D=0.5165;    %steady state duty ratio
L=0.0169; %H
C=1.5273e-5; % F
R=84.4475; % ohms
Ts=5e-5; %20 kHz switching


%SMALL SIGNAL MODEL OF THE BOOST CONVERTER (used for controller design0

ylabes=['Vo Ig '];% lables of outputs
xlabes=['iL Vc '];% state varibles

printsys(As,Bs,Cs,Ds,ulabes,ylabes,xlabes)%prints model of the system

printsys(a,b,c,d ,ulabes,ylabes,xlabes)

disp(['transfer function in s-domain'])
disp(['Vo/d'])% output /input of small signal transfer function
```

```
TFb= zpk(tf(ss(a,b(:,3),c,[0])))% small signal transfer function with d as input

   rlocus (Tfb);
%defining the controller

Kp=0.00090896 ;
Ki=0.22724;


cont = (Ki+Kp*s)/s;

t=feedback (((TFb)*(cont)), 1);

step (t,0.08);
```

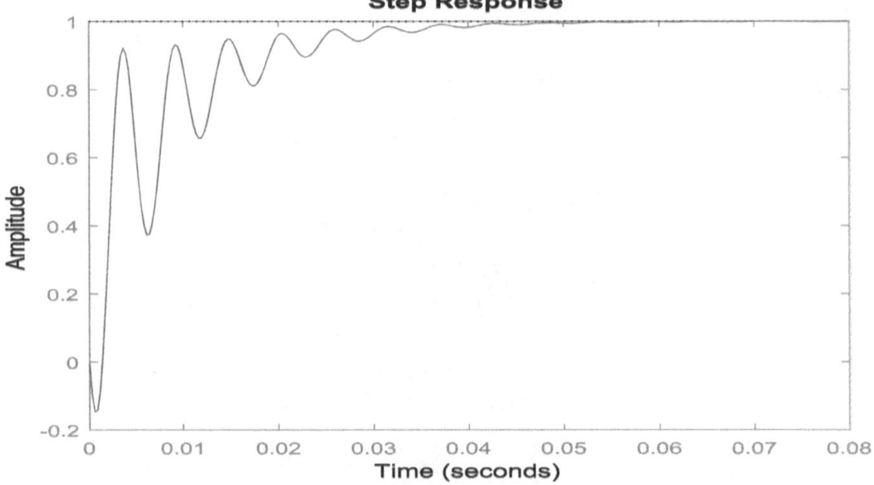

Figure C. 1 closed loop Step response with obtained PI parameters

Appendix D: Direct connection of PV system with a matched load

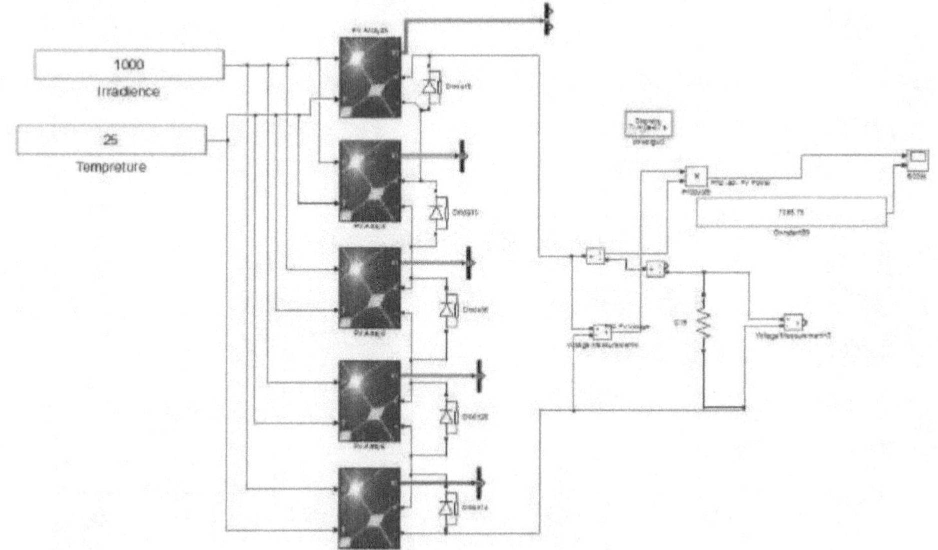

Figure D. 1 Model of direct connection to matched load in simulink

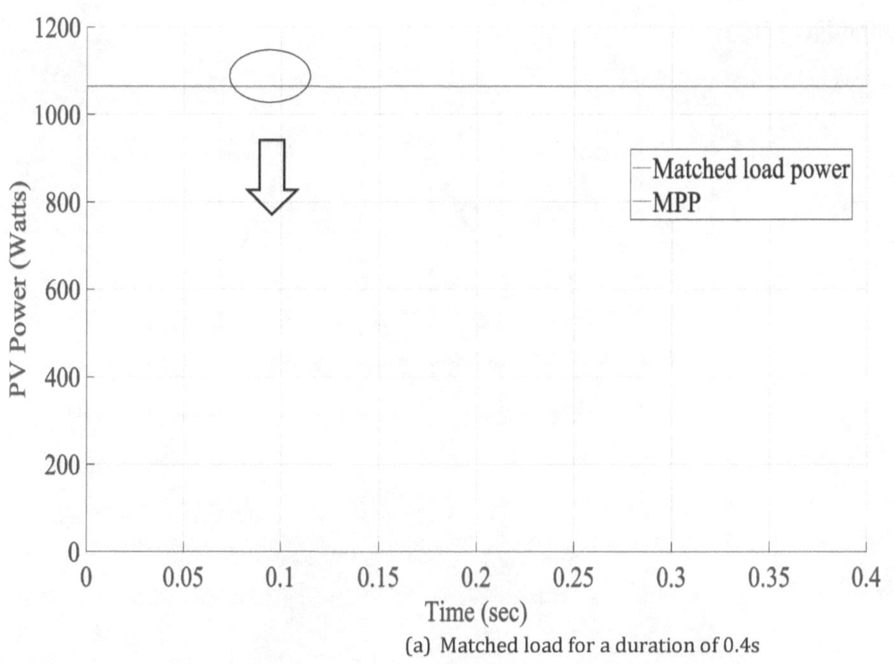

(a) Matched load for a duration of 0.4s

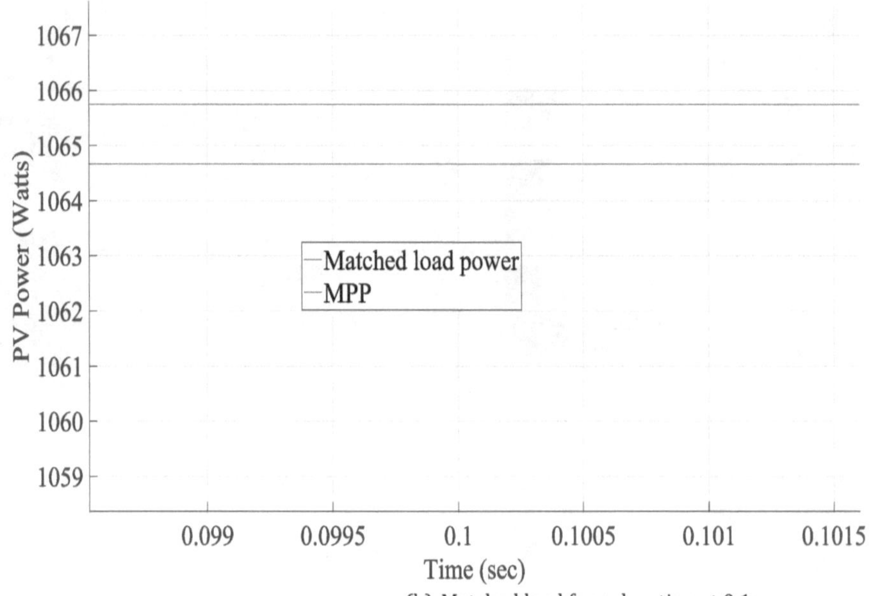

(b) Matched load for a duration at 0.1s

Figure D. 2 Power results with matched load

Appendix E: Fuzzy Logic

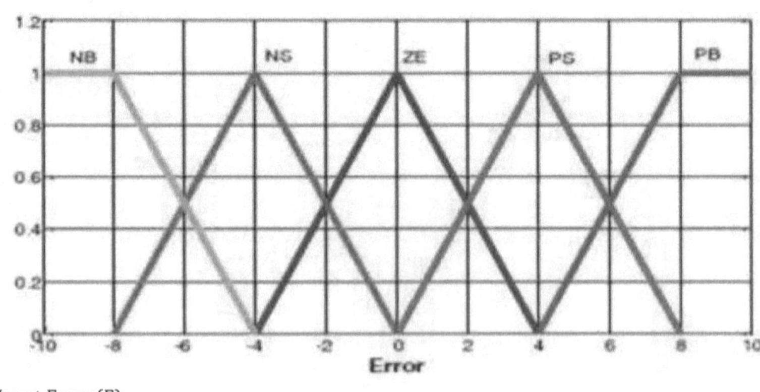

(a) Input Error (E)

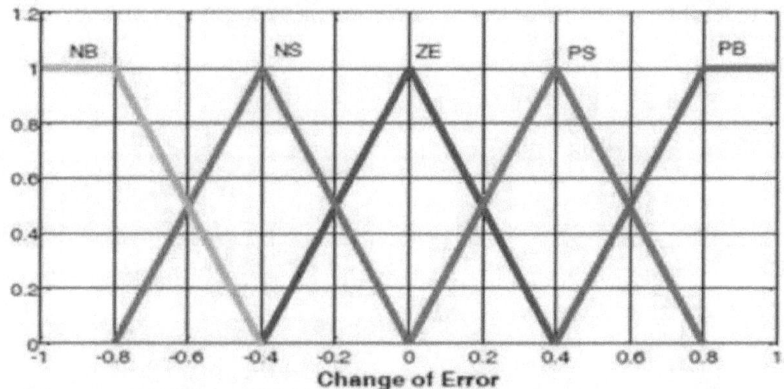

(b) Input change of error (ΔE)

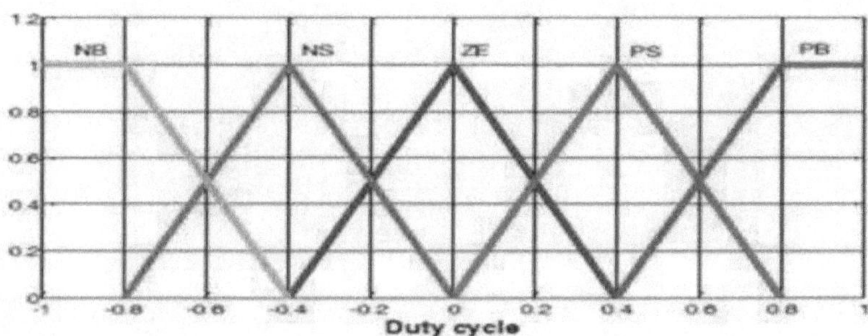

(c) Output change in duty cycle (ΔD)

Figure E. 1 Fuzzy Logic Setup

Appendix F: Results of MPPT of the algorithms for 800,600 and 500 W/m2

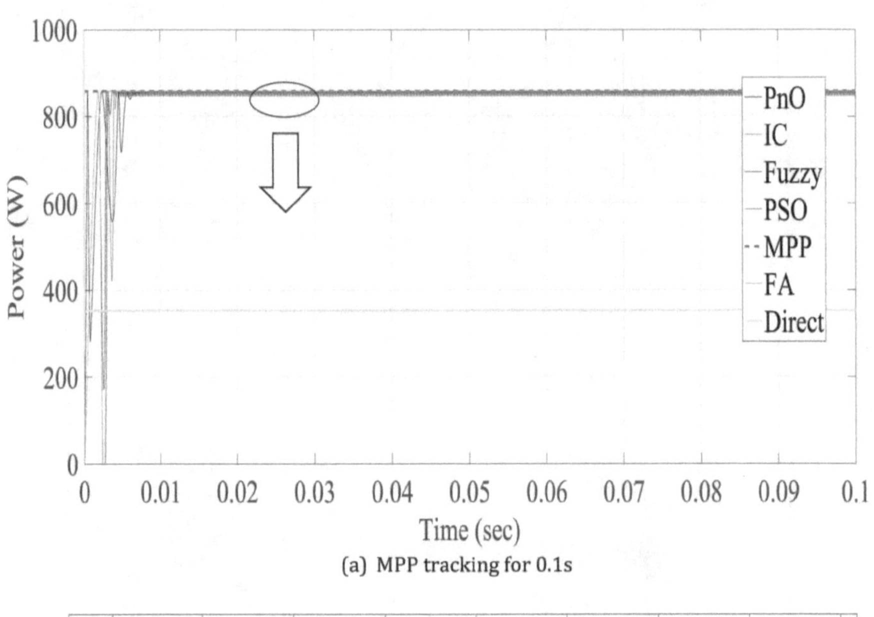

(a) MPP tracking for 0.1s

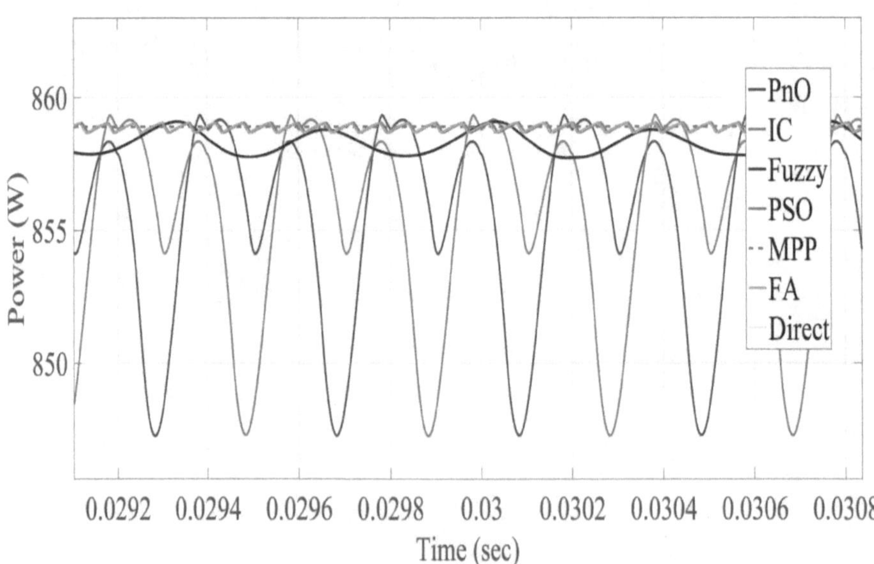

(b)MPP tracking at 0.03s

Figure F. 1 MPPT reults for static irradience of 800 W/m2

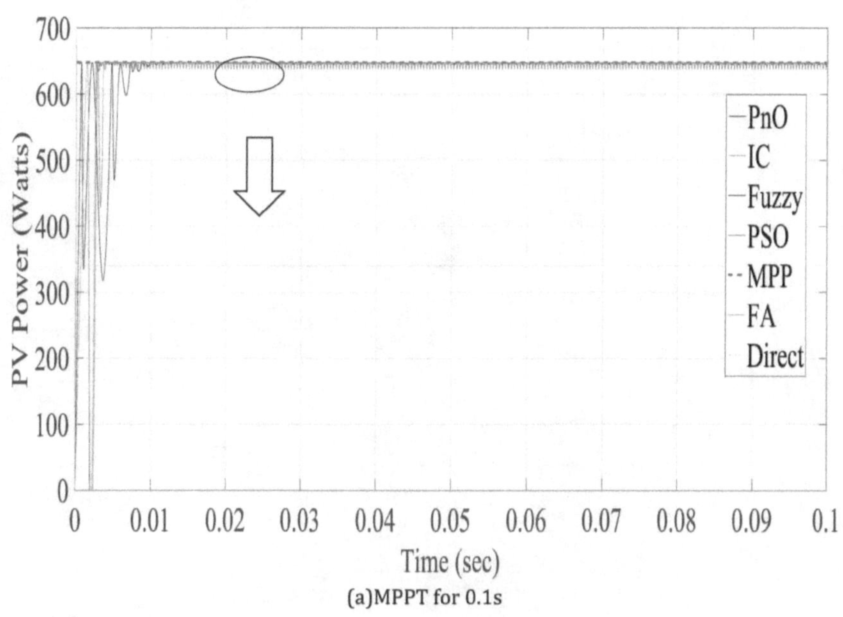

(a)MPPT for 0.1s

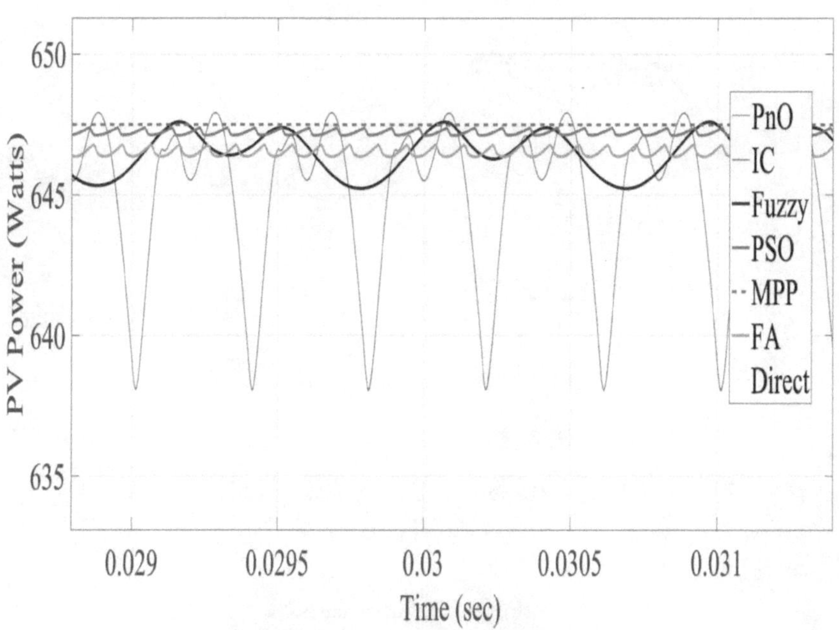

(b) MPP tracking at 0.03s

Figure F. 2 MPPT results for static irradience at 600 W/m2

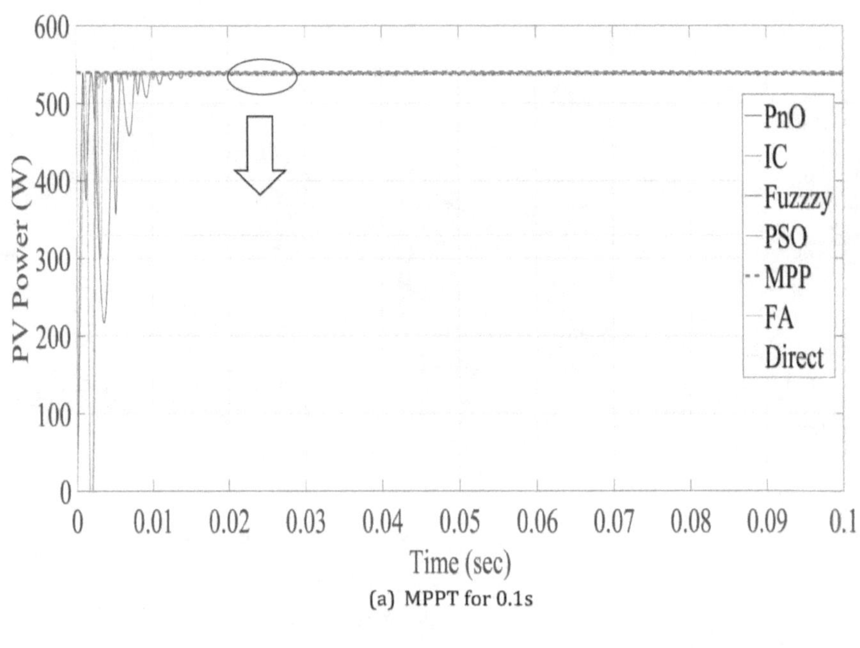

(a) MPPT for 0.1s

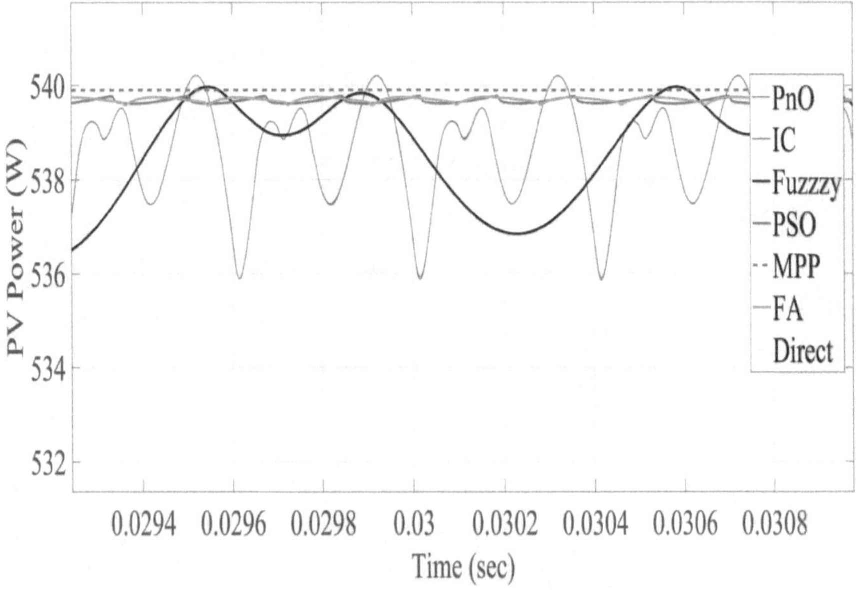

(b) MPP tracking at 0.03s

Figure F. 3 Power results for static irradience of 500 W/m2

Appendix G: Voltage and Current for varying irradiance

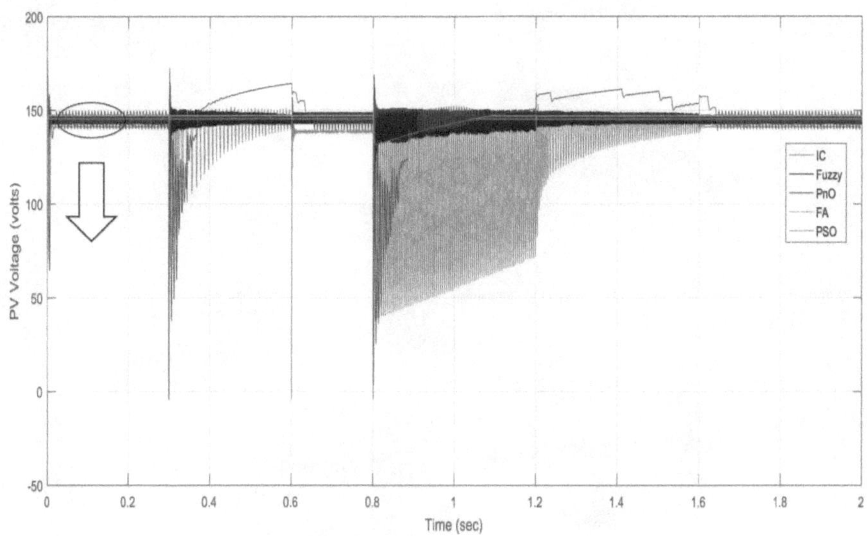

(a) Regulated PV Voltage for 2s duration

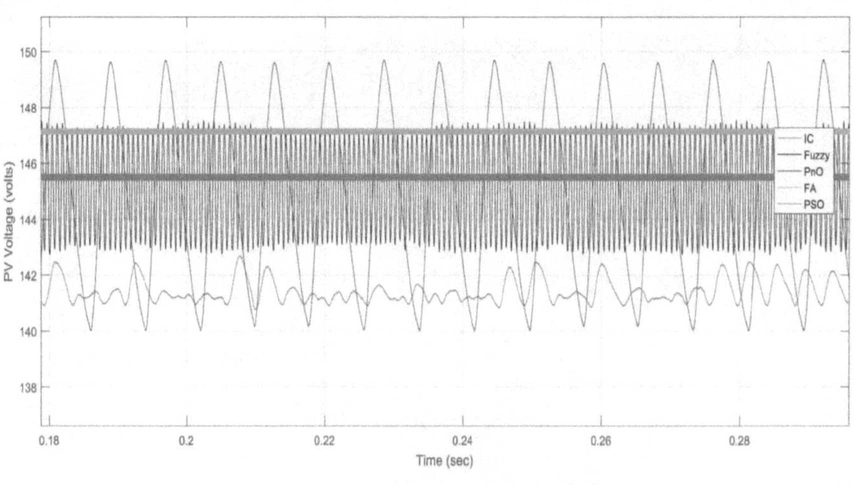

(b) Regulated PV Voltage at 0.2s

Figure G. 1 Voltage for varying irradience

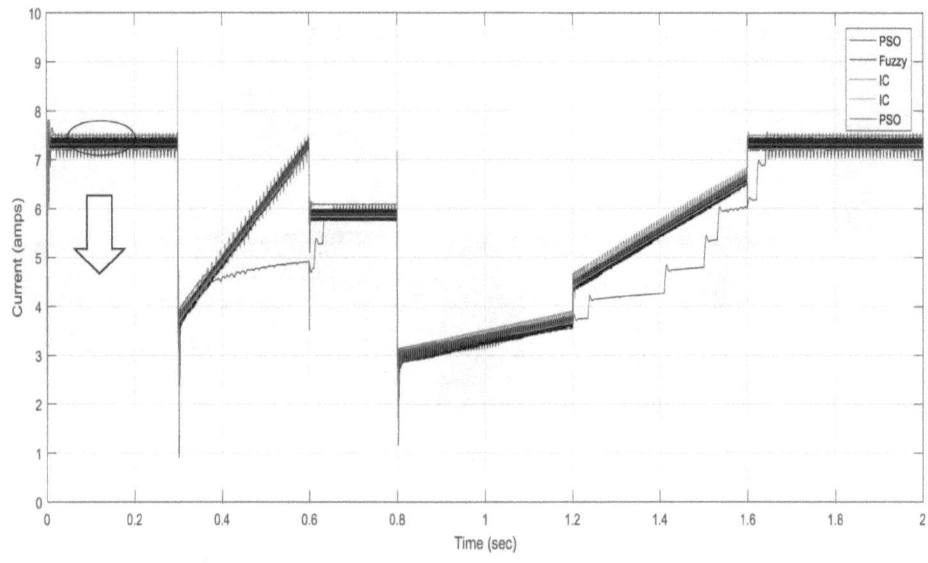

(a) PV current for 2s duration

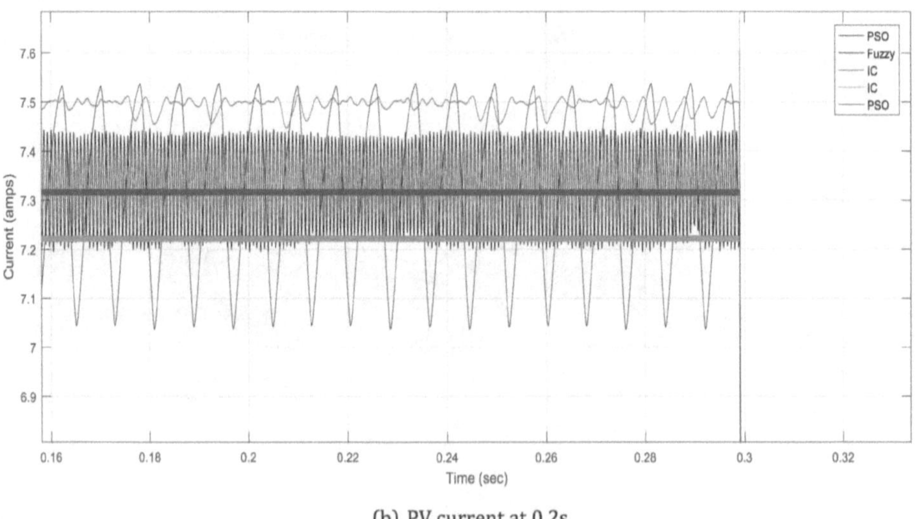

(b) PV current at 0.2s

Figure G. 2 Current for vaying irradience

Appendix H   I-V Partial shading patterns

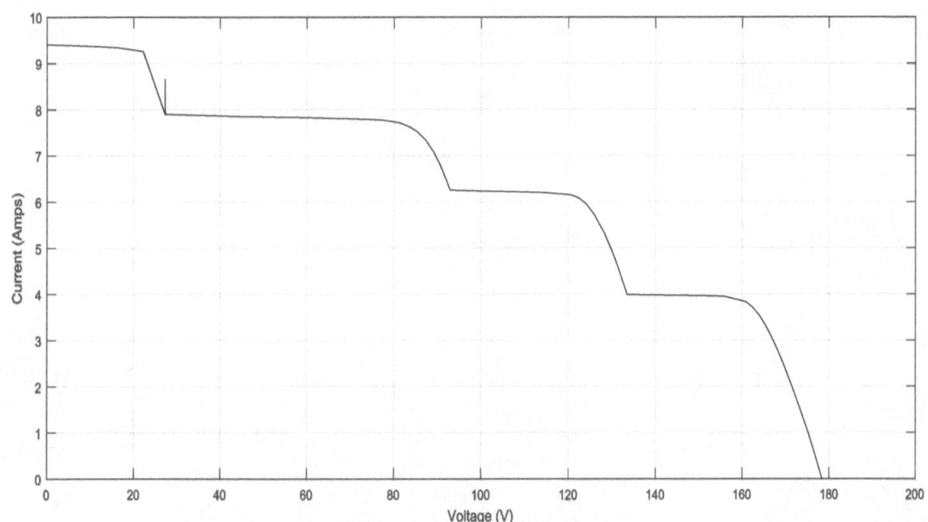

Figure H. 1 I-V curve of pattern 1

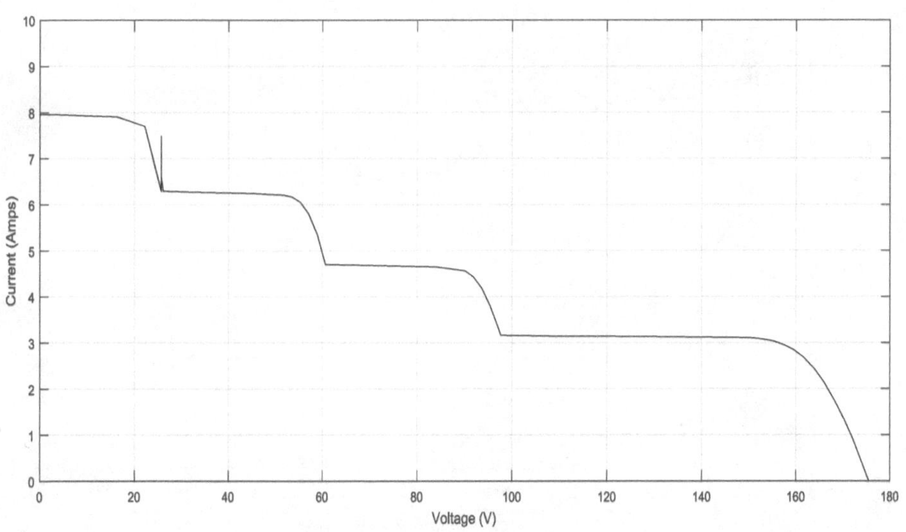

Figure H. 2 I-V curve of pattern 2

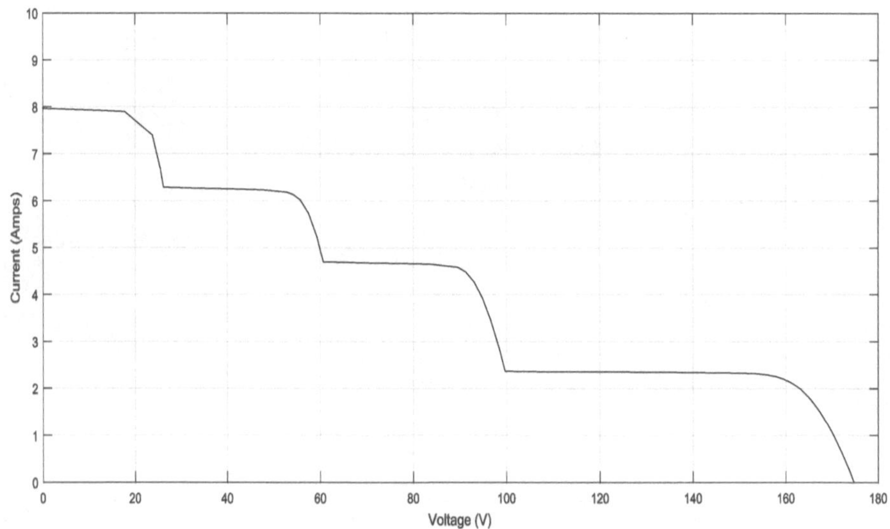

Figure H. 3 I-V cureve for pattern 3

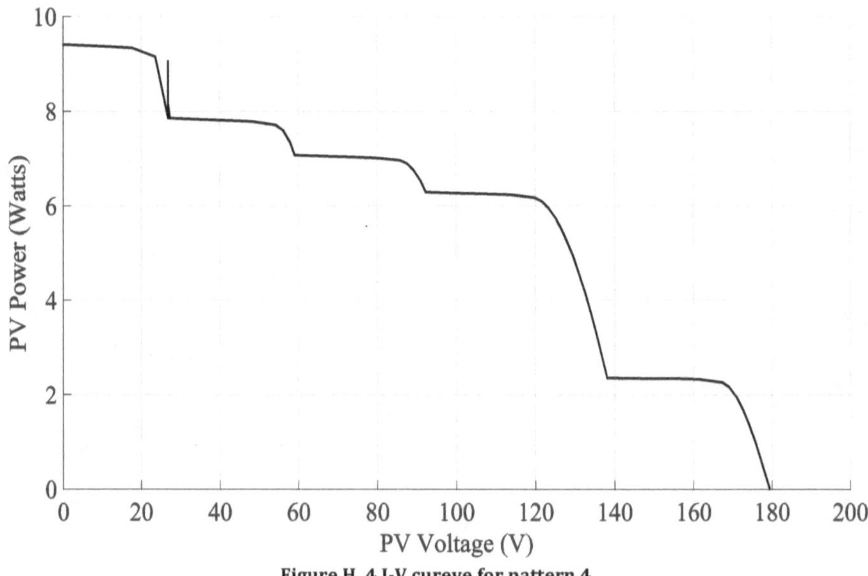

Figure H. 4 I-V cureve for pattern 4

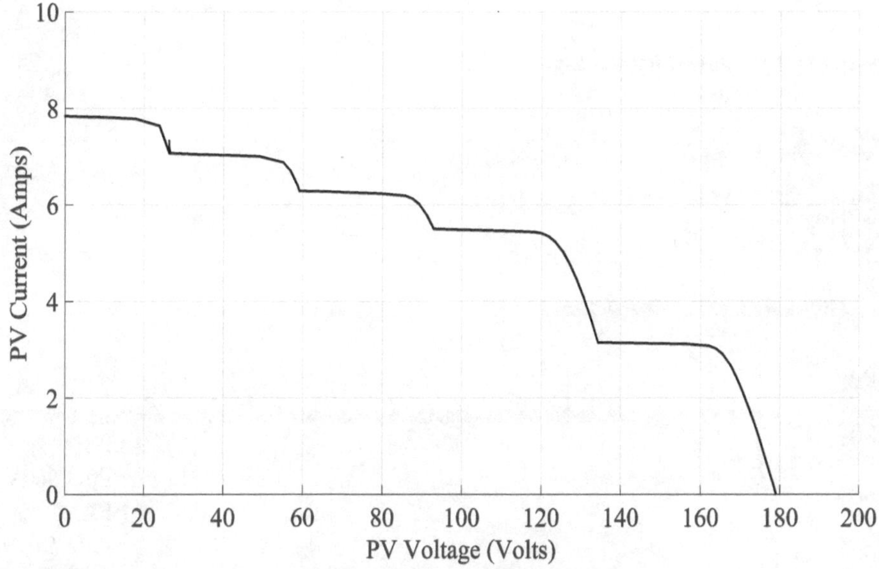

Figure H. 5 I-V cureve for pattern 5

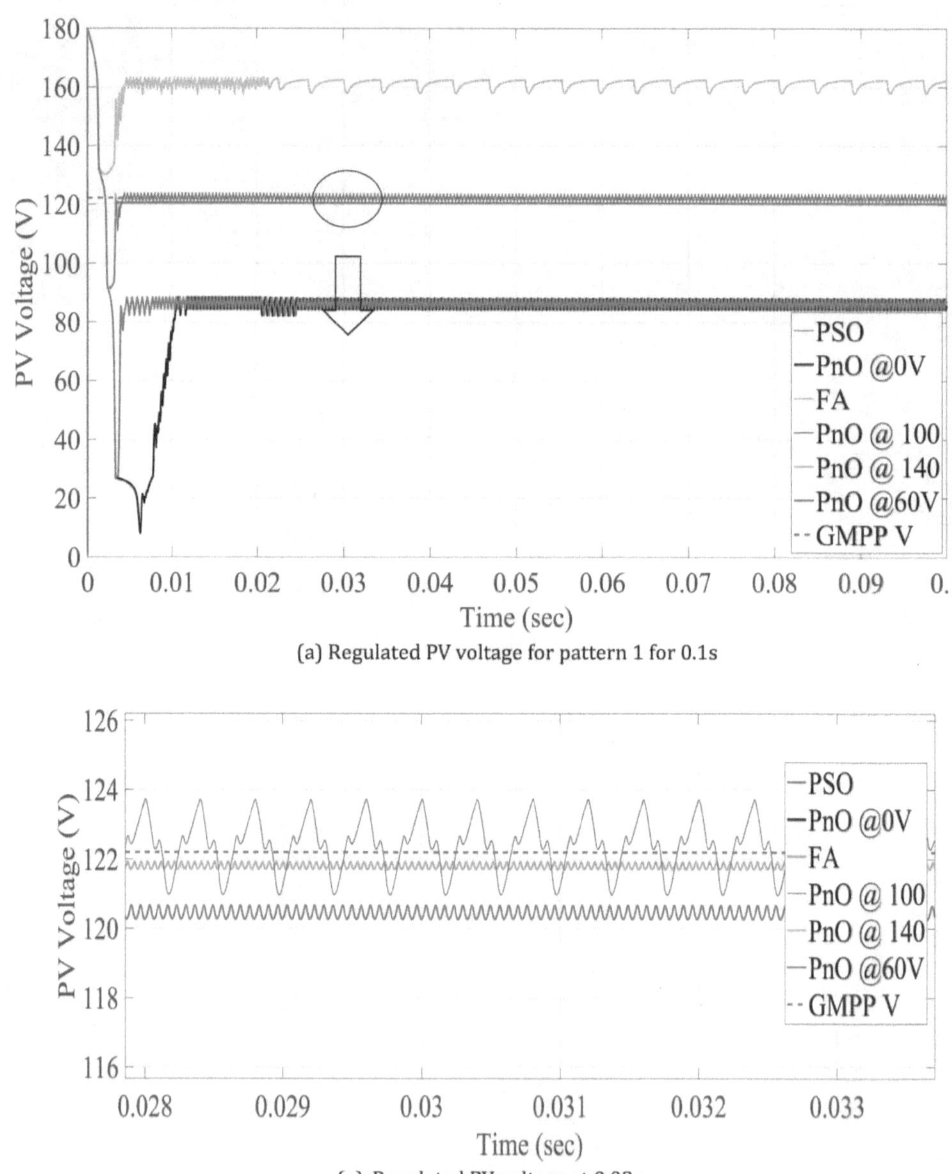

(a) Regulated PV voltage for pattern 1 for 0.1s

(a) Regulated PV voltage at 0.03s

Figure I. 1 PV voltage for pattern `1

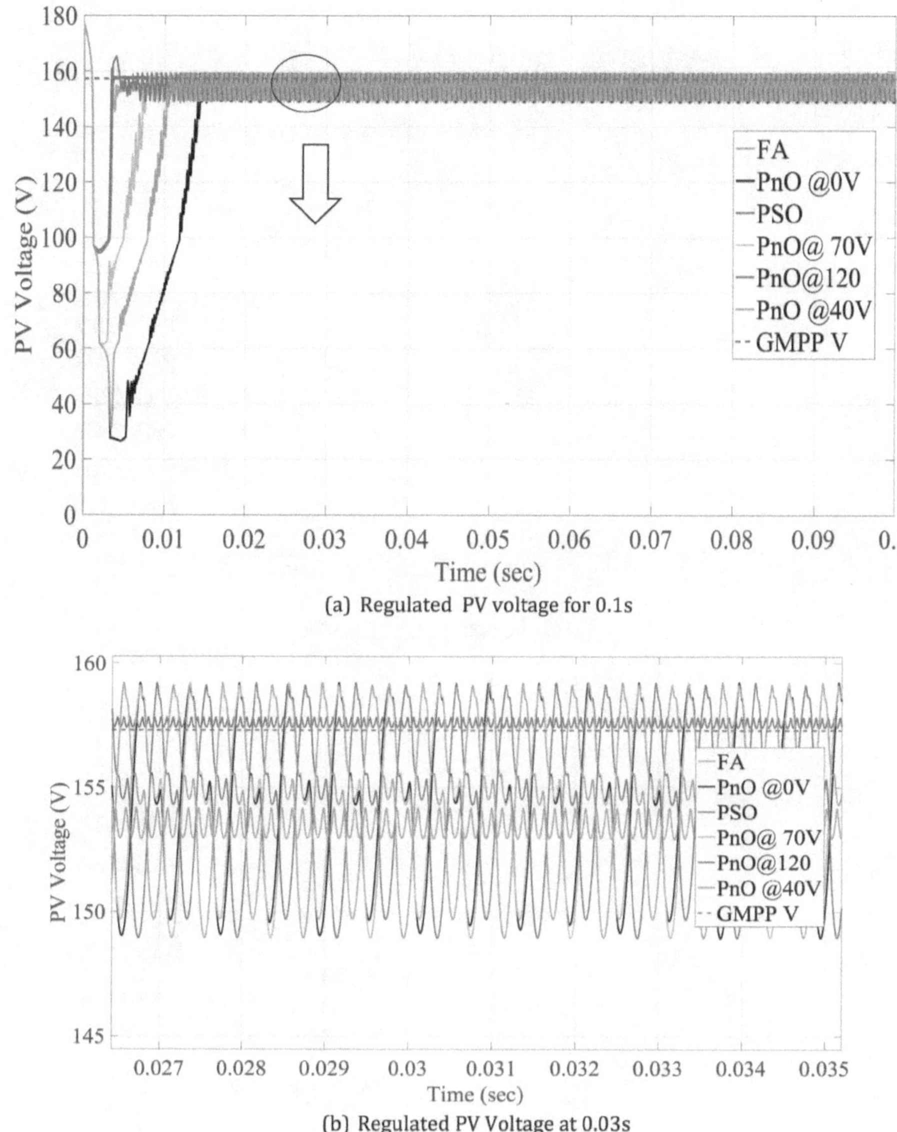

(a) Regulated PV voltage for 0.1s

(b) Regulated PV Voltage at 0.03s

Figure I. 2 PV voltage for pattern 2

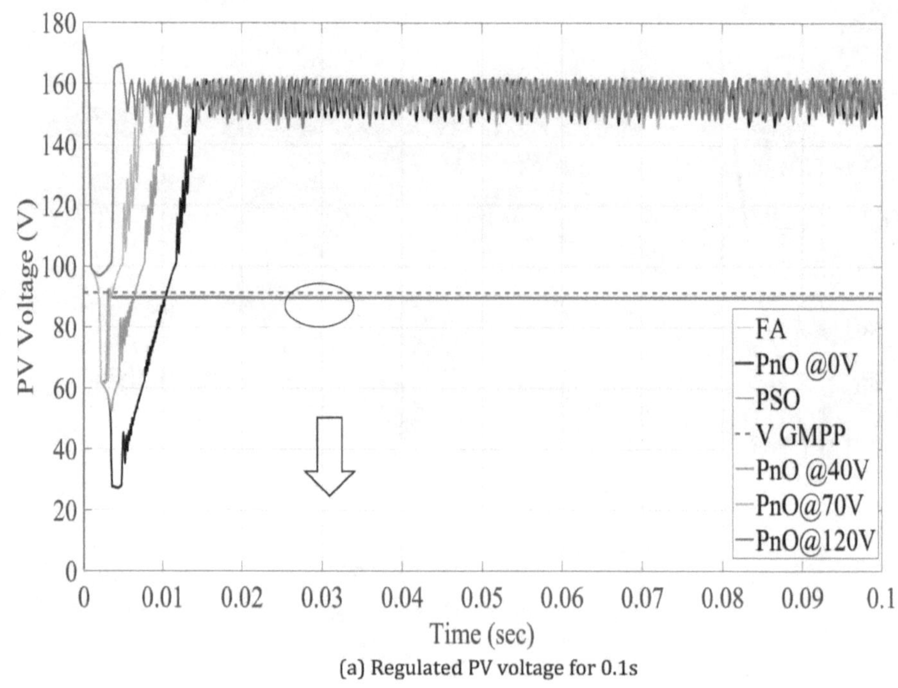

(a) Regulated PV voltage for 0.1s

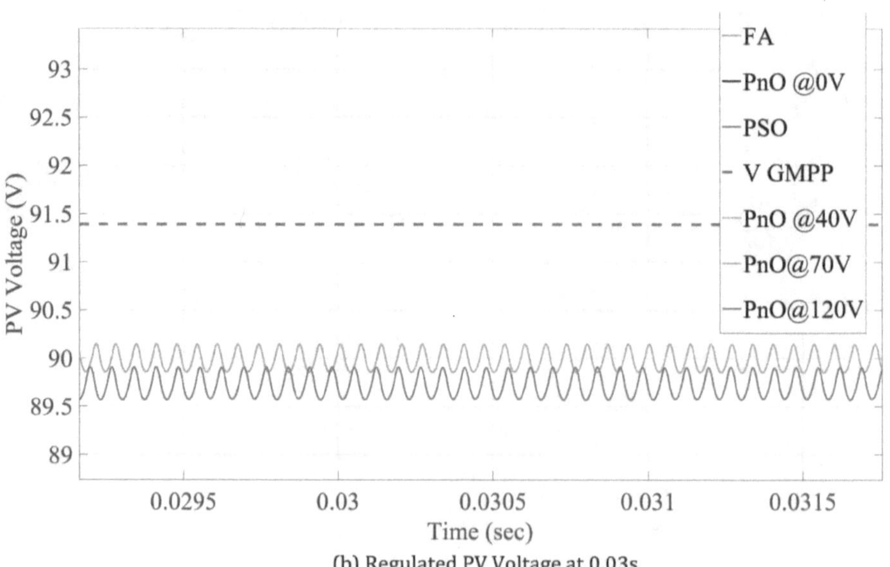

(b) Regulated PV Voltage at 0.03s

Figure I. 3 PV voltage for pattern 3

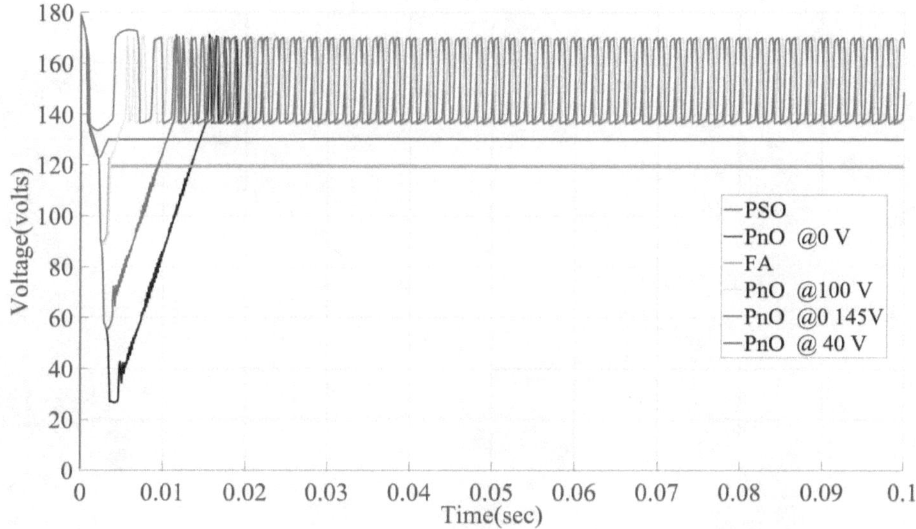

Figure I. 4 PV voltage for pattern 4

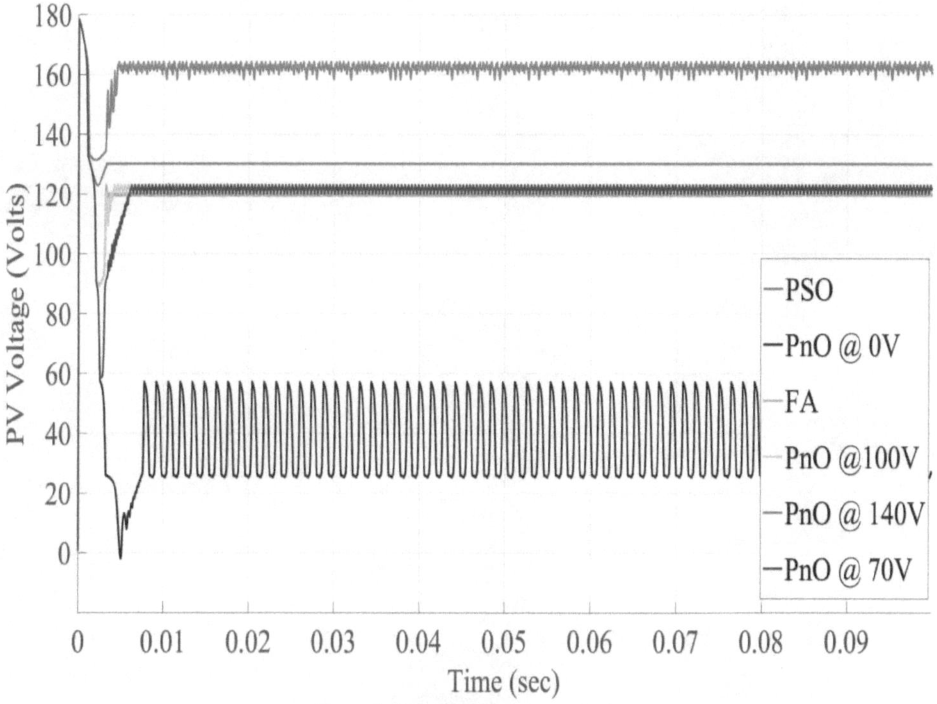

Figure I. 5 PV voltage for pattern5